Human Evolution

Human Evolution

Our Brains and Behavior

ROBIN DUNBAR

OXFORD
UNIVERSITY PRESS

Oxford University Press is a department of the University of Oxford. It furthers the University's objective of excellence in research, scholarship, and education by publishing worldwide. Oxford is a registered trade mark of Oxford University Press in the UK and certain other countries.

Published in the United States of America by Oxford University Press
198 Madison Avenue, New York, NY 10016, United States of America.

Library of Congress Cataloging-in-Publication Data

Names: Dunbar, R. I. M. (Robin Ian MacDonald), 1947– author.
Title: Human evolution : our brains and behavior / Robin Dunbar.
Description: New York : Oxford University Press, 2016. | Includes bibliographical references and index.
Identifiers: LCCN 2016009401 | ISBN 978-0-19-061678-6 (hardback)
Subjects: LCSH: Evolutionary psychology. | BISAC: SCIENCE / Life Sciences /Evolution.
Classification: LCC BF698.95 .D85 2016 | DDC 155.7—dc23 LC record available at
https://lccn.loc.gov/2016009401

10 9 8 7 6 5 4 3 2 1
Printed by Sheridan Books, Inc., USA

For
Zaila, Jared and Arran

Contents

ACKNOWLEDGEMENTS

Much of the research on which this book is based was carried out as part of the British Academy's Centenary Research Project *Lucy to Language: The Archaeology of the Social Brain*. Important aspects of the story are based on allied research projects funded by the Leverhulme Trust, the Engineering and Physical Sciences Research Council and the Economic and Social Research Council, two separate European Union research grants (the *SocialNets* and the *ICTe-Collective* projects) and, last but not least, an Advanced Grant from the European Research Council. I am extremely grateful to all my colleagues on these projects, without whose contributions this book would have been impossible. Ellie Pearce prepared the maps and Figures 1.1, 1.2 and 6.5. The line drawings are by Arran Dunbar.

What We Have to Explain

The story of human evolution has fascinated us like no other: we seem to have an insatiable curiosity about who we are and where we have come from. Conventionally, this story has always been told in terms of the bones and stones that make up the archaeological record – for the very good reason that this is often all we have to go on with any certainty. For the last half century, archaeologists have been loath to stray far from the 'hard evidence' lest they be accused of being speculative. Yet the stones and the bones skirt around what is perhaps the real story of human evolution, namely the social and the cognitive changes that, step by slow and uncertain step, gave rise to modern humans. It is here that the really big questions lie: What is it to be human (as opposed to being an ape)? And how did we come to be that way?

We *are* great apes, since we share with this family most of our biological, genetic and ecological traits. By current consensus, the rest of this family consists of two species of chimpanzees (members of the genus, or biological family, *Pan*), two (maybe four) species of gorillas (genus *Gorilla*) and two (maybe three) species of orang utans (genus *Pongo*). Of these, only the orangs live outside Africa: they are now confined exclusively to the southeast Asian islands of Borneo and Sumatra, although until the end of the last Ice Age some 10,000 years ago, they were widespread in Indo-China and the southern parts of the Chinese mainland.

Until about 1980, the conventional view of our relationship with the great apes was that we undoubtedly belonged to the great ape family, but we and our ancestor species formed a separate subfamily from the other apes. This is because we differ from them in a number of very tangible ways: we walk upright while they walk on all fours, they have large brains by primate standards but ours are so much larger still, we have culture but they merely behave. This picture suggested an early division in the ape ancestry between the lineage leading to modern humans and the one leading to all the other great apes. Since the orang utan lineage can be documented in the fossil record back to around 16 million years ago, it followed that our common ancestor with the apes must have been at least this old.

However, the story changed dramatically during the 1980s when it became possible to look at the genetic (as opposed to purely anatomical) similarities between different species. It soon became apparent that humans are in fact genetically more closely related to the chimpanzees than either of us is to any of the other great apes, though the gorillas come a close second. It is the orang utan – isolated in Asia some 16 million years ago – that is the odd one out. The several species of living African apes (humans, gorillas and chimpanzees) form a single lineage that didn't split up until a much more modest 6–8 million years ago (Fig. 1.1). We belong not to a separate subfamily within the great apes family, but to the subfamily of African great apes. Because we share a common ancestor with the chimpanzees, it makes the chimpanzees (as opposed to some kind of generic early Miocene great ape) the proper comparison for the human lineage, and in many ways the best model for early members of the human lineage (the australopithecines and their immediate predecessors).

So to set the story into context, let me briefly recap the evolutionary history of the African great ape family and our place within it. With this as our background, I will then sketch out five major evolutionary phases that characterize our lineage after its separation from the other African great apes. These phases, or transitions, will be the framework around which I will construct my story of human evolution.

The story so far

The living great apes (including the orang) are the descendants of a dramatic flowering of ape species during the early Miocene era beginning 20 million or so years ago, first in Africa and later in Europe and Asia (Fig.1. 1). Around 10 million years ago, a progressive drying of the climate dramatically reduced the great tropical forests that had been home to this rich diversity of apes during the Miocene. Dozens of species of apes died out, and were replaced by the much more adaptable monkeys who, until then, had been bit-part players on the African and Asian primate scene. One African ape lineage, however, survived and became the common ancestor of the living African great apes. Then, around 8 million years ago, what was to become the gorilla lineage split off. Some 2 million years later, the lineage that eventually

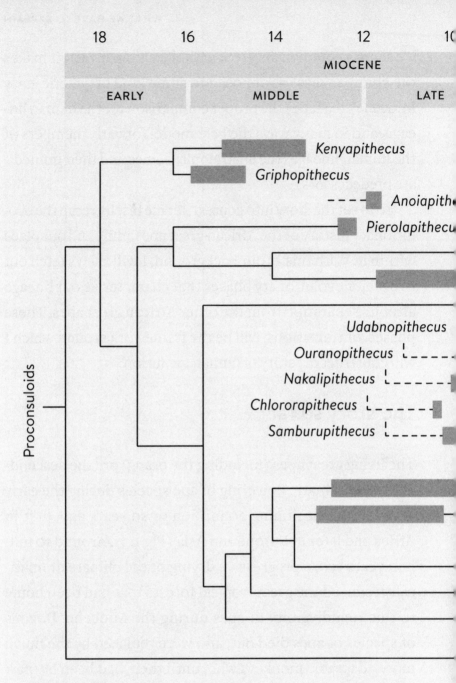

Figure 1.1

The family tree of the ape family, plotted against the geological periods when they lived. The Miocene had a very rich ape community in Africa and Eurasia (the Proconsuloids and their descendants), but most of these died out as the great tropical forests contracted

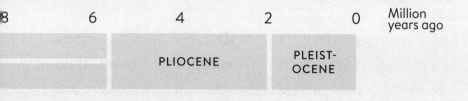

Million years ago

PLIOCENE   PLEIST-
OCENE

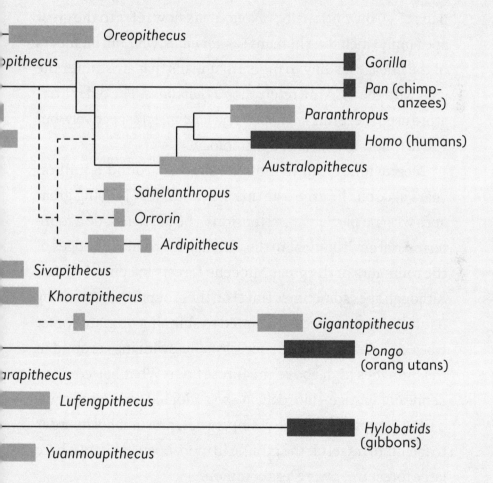

Oreopithecus

...opithecus — Gorilla

Pan (chimp-
anzees)

Paranthropus

Homo (humans)

Australopithecus

Sahelanthropus

Orrorin

Ardipithecus

Sivapithecus

Khoratpithecus

Gigantopithecus

Pongo
(orang utans)

...rapithecus

Lufengpithecus

Hylobatids
(gibbons)

Yuanmoupithecus

towards the end of the Miocene. Dark boxes indicate living genera, light grey boxes extinct ones. Dotted lines indicate relationships that are uncertain (such as the position of the earliest hominins, *Sahelanthropus*, *Orrorin* and *Ardipithecus*, in relation to the human family).

After Harrison (2010).

gave rise to modern humans diverged from the human–chimpanzee ancestor (usually known as the Last Common Ancestor, or LCA) and set off on its own evolutionary trajectory. Much later still, around 2 million years ago – at about the same time as the genus *Homo* made its first appearance in eastern Africa – the chimpanzee lineage split into two, giving rise to the common chimpanzee and the bonobo (or pygmy chimpanzee). Conventionally, taxonomists now refer to the great ape family (including humans) as *hominids*, while all members of the lineage leading to modern humans that arose after the split with the LCA are referred to as *hominins*. The older literature used the terms *hominoids* and *hominids*, respectively, but I shall follow the current terminology.

Meanwhile, from its humble origins around 6 million years ago, our lineage – at this stage still just jobbing great apes with no particular pretensions – began to invade a more terrestrial environment in the woodlands around the edges of the remnants of the great Miocene forests in central Africa. Although apes sometimes travel on the ground, they are all, by nature, arboreal species, adapted to shinning up giant forest trees and clambering (or, occasionally, swinging) around in the branches high above the forest floor. What has come to define our lineage – bipedalism – was adopted early on after we parted company with the chimpanzees, presumably in order to facilitate travel on the ground in more open habitats where large forest trees were less common.

We humans are bipedal apes, and palaeoanthropologists use the anatomical signs of bipedalism to identify our earliest ancestors. Currently, the earliest putative fossil hominin is *Sahelanthropus tchadensis*, although there are doubts as to whether it really is a hominin or 'just another ape'. The find – a near-complete skull found in what is now the Djurab desert on the southern edge of the Sahara in Chad, West Africa – is remarkable both for its age (it is dated to around 7 million years ago, so very close to the LCA) and for the fact that it was found many thousands of kilometres from the nearest other finds of early hominins in East Africa, and almost as far north of the nearest contemporary ape populations in West Africa (indicating that the forests and woodlands once extended much further northwards into what is now the Sahara desert). While some palaeoanthropologists have argued that it is just an ape, others have insisted that the position of its foramen magnum (the hole in the skull through which the spinal cord passes) indicates a bipedal stance, thereby entitling it to be classified as a hominin. Inevitably, perhaps, specimens that date close to the split between apes and hominins are likely to be ambiguous, and so will be difficult to place with any certainty.

The next earliest known hominin fossil is *Orrorin tugenensis*. It dates to around 6 million years ago and was found in the Tugen Hills in Kenya, East Africa. In contrast to *Sahelanthropus*, *Orrorin* consists mainly of limb bones, a jaw and several teeth. The angles of the thigh bone and hip joint[1] suggest fairly uncontroversially that *Orrorin* was bipedal, although it was clearly still a competent climber. In this respect, *Orrorin* seems to bear many similarities to the australopithecines, who became abundant in eastern and southern Africa about a million years later, making it a plausible candidate to be a very early member of the hominin family. Then, from around 4.5 million years ago, the number of fossil finds increases dramatically, reflecting a period when the hominin lineage repeatedly split into ever greater numbers of new species. The australopithecines proper had arrived. At times, there may have been as many as half a dozen australopithecine species alive at the same time, albeit mostly in different parts of Africa (Fig. 1.2).

The australopithecines were extremely successful, radiating over much of sub-Saharan Africa. Yet despite the fanfare associated with their discovery and early recognition as ancestral members of our lineage, they remain little more than bipedal apes. The australopithecines did not differ from the modern chimpanzees in terms of brain size. Like them, they were probably frugivores (fruit-eaters) who may have taken a little meat when they could get it. It is possible that they developed stone tools during the later stages (mostly associated with *Homo habilis* ('handy man'), now considered to be a late transitional australopith), but at best these tools were quite primitive – much like the hammer stones used by chimpanzees today in West Africa.

The next million and a half years from about 1.8 million years ago are dominated by a single species, *Homo erectus*, perhaps the longest lived of all the hominin species. Strictly speaking, it is what biologists call a *chronospecies*, a species that changes through time – not too surprising given its remarkable longevity. It has an earlier phase (*Homo ergaster* and allies) more or less confined to Africa and a later, larger-brained phase (*Homo erectus* in the strict sense) mainly confined to Eurasia. This phase of hominin evolution marks the first expansion out of Africa and into Eurasia (perhaps around 1.5 million years ago, or even earlier), and the appearance of the first *worked* tools (the Acheulian handaxes, named after the first ones found in 1859 at Saint-Acheul in northern France). What is remarkable about this period is its seeming stability. For close on a million and a half years, there is only a modest increase in brain size, and even less change in the shape and style of its stone tools. This phase is probably unique in hominin history for its stability.

Then, sometime around 500,000 years ago, a new hominin species emerges out of the African *ergaster/erectus* stock, eventually taking shape as the first archaic humans, *Homo heidelbergensis*. They mark the beginning of a dramatic explosion in brain size and in the diversity of the material culture. Once again, there are some intermediate populations that form a bridge between *H. ergaster* and *H. heidelbergensis*, but the details are of marginal significance. Meanwhile, the *ergaster* populations disappeared from Africa and Europe as they were replaced by archaic humans. In eastern Asia, however, *erectus* populations survived until as late as 60,000 years ago, and in some diminutive forms (the so-called hobbit, *Homo floresiensis*) on the islands of the Indonesian archipelago until as recently as 12,000 years ago – a mere yesterday in geological time scales.

Of more importance is the fact that archaic humans effected a second major wave of invasions into Europe and western Asia, invasions that eventually gave rise to the archetypal European specialists, the Neanderthals (*Homo neanderthalensis*). Neanderthals developed a particular body form that was especially well adapted to high-latitude habitats, and in particular the cold climates that began to engulf Europe and northern Asia as the Ice Age gathered momentum. Their

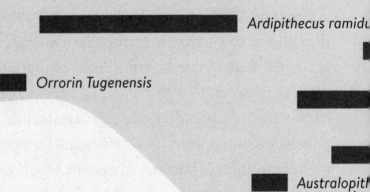

PLIOCENE

Ardipithecus ramidu

Orrorin Tugenensis

Australopitl anamensis

Kenyanthropus platyops

Figure 1.2

Six million years of human evolution, showing the main species and their time spans. Throughout much of our history, there have been several species of hominins alive at the same time. The earliest species were all members of the australopithecine family, which split around 2 million years ago (MYA) into the robust australopithecines and the lineage that led to modern humans.

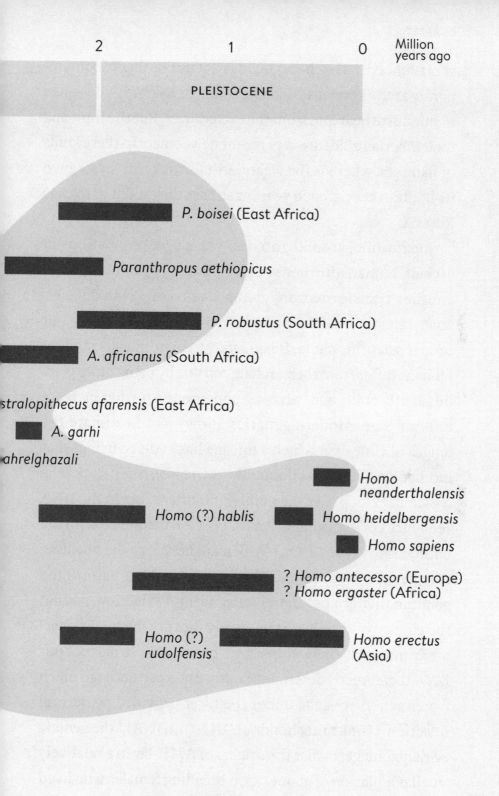

2 1 0 Million years ago

PLEISTOCENE

P. boisei (East Africa)

Paranthropus aethiopicus

P. robustus (South Africa)

A. africanus (South Africa)

stralopithecus afarensis (East Africa)

A. garhi

ahrelghazali

Homo neanderthalensis

Homo (?) hablis

Homo heidelbergensis

Homo sapiens

? Homo antecessor (Europe)
? Homo ergaster (Africa)

Homo (?) rudolfensis

Homo erectus (Asia)

short-limbed, rather heavy build is not unlike that of contemporary arctic specialists like the Inuit (or Eskimo) – designed to minimize heat loss from the extremities. But the Inuit and their Siberian allies are very recent newcomers to these kinds of habitats, whereas the Neanderthals took this strategy to its limits in the 250,000 years or so that they spent in Ice Age Eurasia.

Meanwhile, around 200,000 years ago, populations of archaic humans further south in Africa began to undergo another transformation, giving rise to our own species, anatomically modern humans (AMH) or, to give us our proper scientific name, *Homo sapiens*. Anatomically modern humans differ from their archaic cousins by being more gracile (lightly built) and in having undergone a further increase in brain size. Modern genetics allows us to estimate the length of time over which a lineage has evolved using what has become known colloquially as the *molecular clock*. The molecular clock uses the number of differences in the DNA sequences of two populations or species, combined with the natural rate of mutation, to calculate how long the two lineages have been separated. The focus is on those parts of the genome that are protected from natural selection, so that the clock works only on the steady rate at which DNA naturally mutates. This is important, because those parts of our DNA that directly determine bodily traits can undergo much faster genetic change under the filter provided by natural selection. Using mitochondrial DNA (mtDNA),[2] the genetic evidence suggests that the origins of AMH lay in a relatively small population of about 5,000 breeding females who lived around 200,000 years ago. This doesn't mean that the entire population at the time consisted of only 5,000 females, but rather that only these 5,000 females have contributed to the genetic make-up of all humans alive today.

What produced this new development is not clear. Conventional wisdom points to climate change as the main driver of speciation (the process whereby new species emerge out of ancestral populations), and the origins of *Homo sapiens* may be no different. However, our species seems to have spread very quickly through Africa, quite rapidly replacing the archaic human populations. How and why they replaced archaics so quickly is rather a mystery, especially given the fact that archaics had successfully occupied Africa (and Europe) for at least 300,000 years by the time modern humans appeared.

Then, around 100,000 years ago, one lineage of modern humans in northeastern Africa began to undergo rapid demographic expansion, and by 70,000 years ago had bridged out of Africa across the Red Sea to colonize the southern coastline of Asia, eventually reaching Australia by 40,000 years ago at latest.[3] Reaching Australia was itself a major achievement, because to do so it was necessary to cross a 90km stretch of deep, open sea between the islands of the Sunda Shelf (modern Indonesia and Borneo) and those of the Sahul Shelf (New Guinea, then connected to mainland Australia):[4] they presumably must have had boats of very significant size. Anatomically modern humans mark an important transition in our story because with them comes culture in a way that had never happened before. The period from about 50,000 years ago marks a veritable sea change in the quality and quantity of weapons, tools, jewellery and artwork of a kind and quality never seen before, not to mention tents, lamps and a host of more substantial gear, including boats.

Modern humans came into contact with Neanderthals for the first time in the Levant as they passed through from Africa on their way to Asia. Indeed, it was probably the presence of the Neanderthals in the Levant that prevented modern humans from entering Europe at this point, forcing them to travel eastwards into Asia along the southern coast of the Arabian peninsula. It is very likely that they came into contact with the remnants of *Homo erectus* populations in eastern Asia. It is even more certain that they came into contact with another archaic human species, the Denisovans, in Asia, since it seems they interbred with them. The Denisovans are known only from a handful of bones from a single cave in the Altai Mountains of southern Siberia dated to 41,000 years ago – a cave that was also occupied at different times by Neanderthals and modern humans. Genetic sequencing of the Denisovan genome suggests that they share a common early ancestry with the Neanderthals, and may represent the end point of an early eastwards expansion of the archaic human populations that predated the Neanderthals.

Meanwhile, back in Europe, the archaic human populations gradually became more adapted to the cold northern climate and evolved into the Neanderthals. The Neanderthals remained the undisputed masters of Europe from around 250,000 years ago until 40,000 years ago, when modern humans appeared, like all subsequent historical invaders, on Europe's eastern fringes out of the Russian steppes, perhaps reaching western Europe only around 32,000 years ago. The two species co-existed until about 28,000 years ago, when the last of the Neanderthal populations died out in the Iberian peninsula. The Neanderthals are perhaps the iconic mystery of human evolution. Because they are so close to us both genetically and in time, and died out only relatively recently, we have always had a fascination for them. Why did they go extinct, when they were obviously so well adapted to life in northern climates and had survived in Europe for a great deal longer than our species, anatomically modern humans, has existed? I will return to this question later in the book.

Why we are not just great apes

First, let me return to the central question of this book: we share with the other great apes a long history, a largely common genetic heritage, a similar physiology, advanced cognitive abilities that permit cultural learning and exchange, and a gathering and hunting way of life. And yet we are not just great apes. There are some radical differences. The least interesting of these, although the ones that almost everyone has focused on, are the anatomical differences, and in particular our upright bipedal stance. In fact, most of these traits are just bits of early remodelling to allow a mode of travel that became a route out of certain extinction as the Miocene climate deteriorated and the tropical forests retreated. Much of the rest of the debate has hinged around instrumental behaviours like tool-making and tool use. But in reality these are cognitively relatively small beer – even crows make and use tools, with a brain that is a fraction the size of a chimpanzee's. The substantive difference lies in our cognition, and what we can do *inside our minds*. It is this that has given us Culture with a capital 'C', culture that produces literature and art.

Over the last two decades, a great deal of research has been done – and even more ink spilled in learned journals – arguing the case for culture in animals, and especially in the great apes. The field has even coined a name for itself: pan-thropology, the anthropology of *Pan*, the chimpanzee.[5] It should come as no surprise that behaviours and cognitive abilities that characterize modern humans are also found in some form in our nearest relatives. That is in the nature of evolutionary processes: traits seldom arise completely *de novo* out of the blue. In most cases, they arise as adaptations of existing traits, which become exaggerated or radically modified under the influence of novel selection pressures. We shall examine some of these later. For the moment, the important point to establish is that, yes, humans and chimpanzees share the ability to transmit behavioural patterns socially by cultural learning, and, yes, we can reasonably argue for culture in chimpanzees and other great apes, but the reality is that what apes do with their cultural abilities simply pales into insignificance by comparison with what humans do. This is not to belittle what monkeys and apes do, but rather to identify the substantive issue that seems to get overlooked in all the brouhaha and excitement: humans somehow raised the whole game by a great deal more than just a couple of notches. How did they do this, and why?

There are probably two key aspects of culture that stand out as being uniquely human. One is religion and the other is story-telling. There is no other living species, whether ape or crow, that do either of these. They are entirely and genuinely unique to humans. We know they must be unique to humans because both require language for their performance and transmission, and only humans have language of sufficient quality to allow that. What is important about both is that they require us to live in a virtual world, the virtual world of our minds. In both cases, we have to be able to imagine that another world exists that is different to, and separate from, the world we experience on an everyday basis. We have to be able to detach ourselves from the physical world, and mentally step back from it. Only when we can do this are we able to wonder whether the world has to be the way it is and why, or imagine other parallel worlds that might exist, whether these are the fictional worlds of story-telling or para-fictional[6] spirit worlds. These peculiar forms of cognitive activity are not trivial evolutionary by-products, but capacities that play – and have played – a fundamental role in human evolution. We shall see why in later chapters.

There are, in addition, some other aspects of human culture that will prove to be important. One of these is the social performance of music. To be sure, many other species can be said to produce music, including songbirds and whales, to name but the best known. But only humans seem to engage in music as a *social* activity. For birds, music seems to be mainly a mate advertising display. Humans use music as a mechanism for community bonding in a way that seems to be quite unique. In modern societies, we may often sit listening politely to music in concert halls, but in traditional societies music-making, song and dance are almost indistinguishable and play a crucially important role. This is something we will also need to account for.

What underpins all this cultural activity is, of course, our big brains, and this might ultimately be said to be what distinguishes us from the other great apes. To provide us with a framework for the book, the brain volumes of all the main fossil hominin species are plotted in Fig. 1.3.[7] Seen on the grand scale of the last 6 million years, hominin brain size has been on a steady upswing in which brains trebled in size from their ape-like beginnings among the australopithecines to the brains of modern humans. This seems to suggest that there has been continuous upwards pressure for bigger and bigger brains over time. However, this does not necessarily mean that the selection pressure for larger brains has been increasing steadily over time. In fact, the continuous increase over geological time is an illusion, created by pooling specimens from the different species together. Separating the species out gives a pattern that is more suggestive of punctuated equilibria: each new species generates something more akin to a rapid increase or phase shift in brain size when it first appears, and then brain size stabilizes across time.

In the chapters that follow, I will identify five major transitions or phase changes in the course of human evolution that we need to explain and which will provide the road map for our journey. Each is based on a major change in brain size or ecological circumstance. The first of these is the transition from apes into the australopithecines and involves mainly an ecological and anatomical transition with no evidence for any significant change in brain volume or cognition. After this, there are three phases of brain evolution, starting around 2 million years ago. The first is a significant jump in brain size with the appearance of the genus *Homo* around

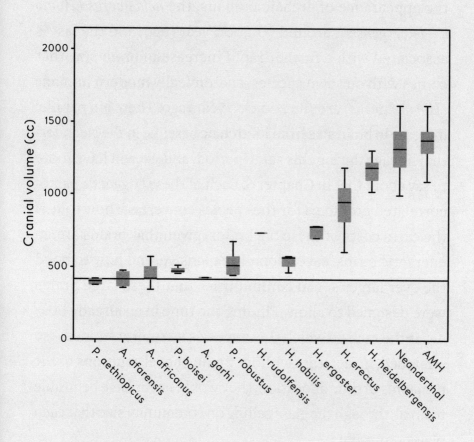

Figure 1.3

Median cranial volumes for major hominin species. The grey boxes show the range within which 50 per cent of all values for each species lie; the whiskers show the range for 95 per cent of the values. The horizontal line marks the equivalent value for modern chimpanzees. Source: De Miguel and Heneberg (2001).

1.8 million years ago (although this is presaged by a smaller, perhaps transitional, increase among the habilines, *Homo (Australopithecus) habilis*). The second is associated with the appearance of archaic humans, the *heidelbergs (Homo heidelbergensis)*, around 500,000 years ago, and the last is associated with a further rapid increase in brain size that comes with our own species, anatomically modern humans (*Homo sapiens*) around 200,000 years ago. There is a parallel increase in brain size from its archaic baseline in the Neanderthal lineage during this same period, and we will have more to say about that in Chapter 6. Each of these triggers a pair of correlated problems for the species concerned: how to fuel the extra costs of these big brains, given that brains are, in energetic terms, exceptionally expensive, and how to bond the ever larger social communities that these large brains were designed to allow. Finding the time in an already busy day is the rack on which these species were all stretched – so stretched, in fact, that without finding novel solutions to the problems of time allocation, they would never have been able to break through the glass ceiling on community size that each phase represents.

To these four fundamental transitions, I will add a fifth that does not actually involve any changes in brain volume: the Neolithic Revolution that took place around 12,000–8,000 years ago in the Near East. The Neolithic is particularly intriguing because it is a reversal of everything that went before. It is characterized by two major innovations: the switch from nomadism to settled villages and, eventually, the invention of agriculture. Although the agricultural revolution has always attracted more attention than anything else, in fact agriculture was just the means to an end. The real revolution was being able to live in settlements: irrespective of why communities gathered together in fixed settlements, doing so created social stresses that had to be neutralized before the Neolithic could take off. Once this problem had been resolved, however, it opened up the possibility for evolving ever larger communities, and hence eventually the rise of city-states and petty kingdoms and the whole historical development that eventually gave rise to the nation-states of the modern world. Understanding how we managed this transition is thus a key part of our journey.

The way ahead

As I observed at the start of this chapter, the archaeologist's bread-and-butter is a combination of stone tools and fossil bones, combined with a bit of local geology. However, this conventional focus on 'stones and bones' has inevitably left the social aspects of human evolution, and even more so their cognitive underpinnings, largely unexplored. The archaeologist's understandable worry has always been that it is just too easy to infer what you will about social behaviour from the sometimes scrappy and invariably indirect evidence available in the deep time record. Yet it is precisely these social and cognitive aspects of our biology that mark out the road that led, tortuously and often uncertainly, from the Last Common Ancestor some 6–8 million years ago to the modern humans that we now represent. If we want to account for that trajectory, we have to grapple with this murky, unseen social world, however difficult that may be.

In the societies of primates (including contemporary humans), the social community takes the form of a highly structured network of individuals linked to each other through ties of kinship, friendship and obligation. The way in which these social networks are structured in terms of kin and non-kin and the way they are distributed in space both have implications for how easy it is for an individual to call for assistance, as well as how well the relationships on which the coherence and persistence of the network depend can be serviced.

We are now in a position to address these kinds of questions because we have a much better understanding of primate social behaviour and ecology. This improved understanding is crucial because it neatly sidesteps one of the perennial problems that have long bedevilled attempts to reconstruct the behaviour of fossil hominins. Hitherto, the standard approach has been to identify a living species that is thought to share some key trait with a particular fossil hominin, and then assume that the fossil had the same ecology and social organization as the living species. At one time or another, chimpanzees, gorillas, baboons, lions, hyenas and even African wild dogs have all had the distinction of being 'the model' for early hominins. It was for this reason that the great fossil hunter Louis Leakey sent Jane Goodall and Dian Fossey off into the forests of central Africa to study chimpanzees and gorillas: he hoped they would be able to tell us something about how fossil hominins might have behaved. These 'analogue' models[8] suffer from the weakness that they are invariably based on a single similarity shared by the living and fossil species – which might or might not actually be relevant to why the living species has the social system it does. They also assume that each species has a characteristic 'style' of behaviour, and while this is very broadly true, the one thing we have learned from the last half-century of field studies on primates is just how adaptable most species are behaviourally and ecologically.

The approach that I will adopt here is very different. It exploits our much improved understanding of how primates allocate time to the various core activities (feeding, travel, rest, social bonding) that are crucial to their ability to survive in a particular habitat. This approach builds on a series of time budget models that we have developed for a number of monkey and ape species – models that allow us to predict exactly how much time an animal should devote to each of these core activities in a particular habitat. The key issue here is that the length of the working day is limited (we all sleep at night) and all these core activities have to be performed within the waking day. The fact that we are dealing with a biological system here is a crucial advantage: in biological systems, changes in one component cannot happen without having knock-on consequences elsewhere in the system. A species cannot increase its brain or body size without affecting the time it has to spend feeding, and it cannot change its feeding time allocation without affecting the time available for other equally crucial activities like travel or socializing. In a word, the numbers have to add up. And this gives us a powerful tool for exploring species' responses to changing circumstances.

The second major foundation for our task is the social brain hypothesis, since this provides what is in effect the fulcrum that will underpin the time budget analyses. Originally proposed by the psychologists Andy Whiten and Dick Byrne as an explanation for the fact that primates have much larger brains for body size than all other mammals, the social brain hypothesis has since come to be an explanation for the correlated differences in cognition and sociality *between* primate species. The crucial feature of this hypothesis is that it provides a quantitative equation relating brain size to social group size. The fact that this relationship is extremely robust and almost independent of any *direct* effects due to ecology means that we have a way of predicting typical social group size in fossil species. This provides us with two key insights into time budgets. One is the fact that, because brain size predicts group size, we can determine how much extra time is needed for bonding larger groups; the other is that an increase in brain size has to be fuelled by additional foraging time. The question we ask for each species is simply this: how did they accommodate these additional demands on their time? And if their time budgets were already stretched to their limit, what novel solutions did they find to make the extra time they needed available?

What we have to explain

How was it, then, that one particular lineage of African great apes set out on a trajectory that appears to have been unique? And, having done so, how was it that one particular subset of these early hominins emerged from the scramble of the australopithecine radiation to colonize the Old World and ultimately become the only member of this adaptive array to survive the climatic changes of the later Pleistocene? And, again, how was it that, of the flurry of highly successful lineages that emerged out of the genus *Homo* in the middle Pleistocene, only one lineage – our own – managed to make it through to the present?

To a large extent, the trajectory that defines our pathway over these 6–8 million years reflects the dramatic changes in brain size and organization that mark out the sequence of events that makes up this story – the speciations, the migrations, the extinctions and the cultural novelties that litter the timeline of hominin evolution. Associated with these changes in brain size are a number of other core traits, some of which we can infer from the archaeological record and some of which we know reliably only from modern humans. I summarize these in Table 1.1 under four main headings. Some are anatomical, some behavioural or cognitive, but all have to be fitted into a single seamless sequence against both the changes in brain size (and hence group size) and the constraints of time, as well as the archaeological record. It is this triangulation between the different sources of information that makes our task possible, since it allows us much less room for speculative manoeuvre than has hitherto been the case. We cannot assemble the pieces of the jigsaw in Table 1.1 in any random order and simply make up some plausible story for the particular pattern we happen to favour. Instead, our approach will allow us to provide principled reasons for assembling the pieces in a *particular* order – or at least arriving at a limited number of alternative possible ways of doing so.

Some of the traits listed in Table 1.1 will be familiar to

ANATOMICAL MARKERS	ARCHAEOLOGICAL MARKERS
bipedality	fire
striding gait	hearths
plantar foot	changes in tool style
pelvic restructuring	decorative art / jewellery
brain size increase	home bases
loss of canines	
delayed dental development	
modern human lifehistory	
handedness	
gracilization	
menopause	
precocial babies	

Table 1.1

Modern human traits that differentiate us from the apes.
Our task is to build a sequence in which these were acquired.

BEHAVIOURAL MARKERS	COGNITIVE MARKERS
fission–fusion sociality	theory of mind (mentalizing)
laughter	high order mentalizing
dietary change [esp. underground storage organs]	
meat-eating	
cooking	
hunting	
grandmothering	
language	
(romantic) pairbonds	
allo-parental care	
division of labour	
story-telling	
music and dance	
religion	

palaeoanthropologists and are the bread-and-butter of traditional accounts of human evolution. These include bipedalism, the changes in pelvic structure and the acquisition of a plantar[9] foot, the loss of canines, the increased gracilization (i.e. lighter body build) of modern humans, the progressive increases in brain size, the acquisition of a modern human lifehistory with delayed maturation (signalled by delayed molar eruption) and precocial babies, tools in all their complexity, hunting and artwork. Others (fission–fusion sociality, division of labour, grandmothering, the menopause, cooking, religion and pairbonding) have played a central role in anthropological discussions of human social evolution but often lack plausible archaeological signatures by which we can recognize them in the fossil record. Some, however, are genuinely novel and have never really been considered in the context of human evolution: these include music and dance, story-telling, religion, those forms of social cognition known as theory of mind or mentalizing, and laughter. These have, I shall argue, played a particularly important role in the story of human evolution. Our task will be to explain why these changes took place, and why they occurred at the particular times and places they did.

This, then, will be an exercise in detection. We have the crime scene before us in the archaeological record, tantalizingly imperfect as every crime scene always is. Our task is to try to infer what happened where, when and why. The social brain hypothesis and the time budget models provide us with a forensic toolkit that we can apply rigorously to each stage in the unfolding story. We shall proceed, like all good detectives, by trying to fit the bits of the jigsaw together. Because our forensic toolkit is quantitative (the numbers must add up in a time budget model), we can't just make the jigsaw pieces fit into some arbitrary pattern that happens to suit our particular predilections. We will build this picture step by step through time, placing the new crises that beset each species in the context of how its predecessor(s) resolved the ones they faced. By working incrementally in this way, we should be able to build a more coherent picture than has hitherto been possible.

There are two notes of caution that I should probably sound.

One is that many palaeoanthropologists will view this whole enterprise with something close to horror, so sacred are the stones-and-bones to their way of thinking. They have a long history of suspicion about new approaches and new technologies: many expressed disbelief when molecular genetics upended hominid (i.e. ape and human) taxonomy in the 1980s. The moral is that, rather than viewing new approaches with suspicion, we must ask how we can use them to help us better understand an all too fragmentary archaeological record. Science proceeds not by getting the answers right on the first shot, but by making us ask questions. In this book I ask entirely new questions about the story of human evolution, and propose entirely new approaches to answering them. I don't doubt that the details of the story laid out here will change as we acquire new fossils and new kinds of technical understanding; that is nothing new in a discipline where, famously, every new fossil is heralded as changing the story of human evolution. The important point is that we ask questions that make us interrogate the archaeological record in new ways.

The second caveat concerns the exact taxonomic status of different hominin fossils. Taxonomy has been the focus of much of the debate in the study of human evolution over the last century, and a great deal of ink has been spilled on this issue. I do not intend to spill any more ink at all on the subject, and many will no doubt be affronted by such a cavalier attitude. This is not to say that taxonomy isn't important. I simply crave indulgence for ignoring it for now on the grounds that we do not, in my opinion, have a sufficient handle on the detail to be able to undertake a more refined analysis. Instead, I want to dispense with the detail and focus on the big picture: how was it that these species managed to survive where they did, and why, eventually, did most of them go extinct. To the extent that this exercise is a success, it will provide us with the justification for further investigations at the level of individual populations – and at that point the details of taxonomy will surely become more important since we will need to know exactly who these individual populations are.

First, however, I need to introduce the key ideas and concepts that will form the basis of our story. The next chapter summarizes some essential principles of primate social evolution that will frame our whole approach. In effect, these provide the rationale for much of what follows. Hominins have to work within this framework precisely because they are primates – or, at least, if they escape it, we shall need to be able to show how, when and why they did. The following chapter will then explore in more detail the two crucial theories that provide the template for the rest of the book. Our whole exploration of human evolution will depend on these two theoretical dimensions.

The Bases of Primate Sociality

Primates are, above all, intensely social, and their peculiar kind of bonded relationships allows them to form groups that remain stable and coherent over time. The main reason that primates live in groups is as a defence against predators. An animal's risk of being caught by a predator increases as it becomes more terrestrial and occupies more open habitats with fewer trees to provide refuges. Under these conditions, group size increases, and to support this their social relationships shift to being ever more intensely bonded, presumably in order to ensure that individuals stick together and come to each other's aid when necessary.

Living in close proximity to other individuals may have its benefits, but it also has its costs. These arise from three separate sources: direct and indirect costs, and freeriding. The direct costs arise from conflicts within the group: altercations between individuals over food or the safest resting sites. All of these inevitably increase in frequency as groups get larger. The indirect costs are the fact that more travel time has to be allocated within the waking day as group size increases in order to allow the group to visit enough food patches to satisfy everyone's nutritional requirements, and that means less time is available for other activities.

Furthermore, travel consumes energy, so there are knock-on consequences for feeding time as more foraging will be needed to pay for the extra energy used in travel. Travel also exposes the group to the additional risk of encountering predators. Last but not least, a primate group is an implicit social contract (it is a collective solution to the problem of predation), and social contracts are always susceptible to being broken by freeriders – those who take the benefit of the contract but don't pay the costs, thereby benefiting doubly at everyone else's expense. In this case, the benefit is collective protection from predation. Individuals who exploit fellow group members impose a burden that these eventually become unwilling to bear. As with the other costs, the risks of being exploited become more acute as group size increases.

In primates, the direct costs are borne mainly by the females, because the stresses of living in a group – the casual jostlings in cramped conditions, the occasional conflicts over access to food or safe refuges – impact on female menstrual cycle endocrinology and lead to infertility. These stresses tend to accumulate down the dominance hierarchy simply because lower-ranking animals have more individuals who can harass them. When the stresses mount up, they block the normal menstrual hormones, and the result is an anovulatory menstrual cycle – a cycle in which ovulation does not occur, and the female loses a chance to conceive. Every time this happens, her lifetime reproductive output (and hence her fitness) is reduced by a fraction. The rate at which this form of temporary infertility accumulates down the dominance hierarchy is sufficiently steep that, in some species, a female ranked tenth in her group can be completely infertile.

As this pressure mounts with group size, females will want to live in smaller groups. Thus female reproductive strategies become a crucial factor that puts a natural brake on runaway increases in group size. When predation pressure favours an increase in group size, primates will have to neutralize these stresses, otherwise groups will inevitably fragment and living in large groups will be impossible. The next section explores how they do this.

Defusing the stresses of group-living

Monkeys and apes neutralize the stresses created by living in groups by forming coalitions that buffer their members against harassment. The intensely bonded relationships that underpin these coalitions seem to involve two separate processes that interact to create these bonds. One is an emotionally intense mechanism triggered by social grooming. Social grooming seems to work because it triggers the release of endorphins in the brain, as was shown in a very neat experiment by the Cambridge neuroscientist Barry Keverne and his colleagues some years ago. This probably allows two animals who groom together to stay together long enough to build up a relationship of trust and obligation, which forms the second cognitive component. Although other neurotransmitters and neuroendocrines such as oxytocin – the so-called 'love hormone' – are also released during close social interactions and play an important role in mammalian sociality, the evidence points to a unique role for endorphins in the maintenance of close relationships in anthropoid primates. Grooming and other forms of light social touch (stroking, cuddling) trigger activation in a special set of neurons (the unmyelinated afferent c-tactile fibres) that respond specifically to light touch on the skin and convey these sensations directly to the brain. Although we don't know if it is these neurons that mediate the endorphin activation in response to grooming in primates, we do know from recent PET scanning experiments that we have been running that light stroking between romantic couples triggers an endorphin response in humans. Psychologically, endorphin release is experienced as a mild opiate 'high', light analgesia, pleasure and calmness. It seems that this is intimately involved in the processes for forming the kind of deep attachments that we find in anthropoid primates (including, of course, ourselves).

Given the importance of grooming in creating and maintaining coalitions, it is perhaps not surprising to find that time devoted to grooming correlates with social group size in Old World monkeys and apes (though not, so far as we know, in New World monkeys or the prosimian primates like galagos and lemurs) (Fig. 2.1).[1] However, this does not necessarily mean that animals groom more individuals in larger groups. In fact, in the most social primates, the number of animals groomed *decreases* as group size and grooming time increases. Both across species and within species, the cliquishness of grooming (i.e. how broken up the group is into small, semi-independent grooming cliques that mostly groom among themselves) increases as group size increases. This seems to be because, as the stresses increase with group size, it becomes more and more essential to ensure that one's coalition works as reliably as possible, and so animals increasingly concentrate their grooming effort on their most

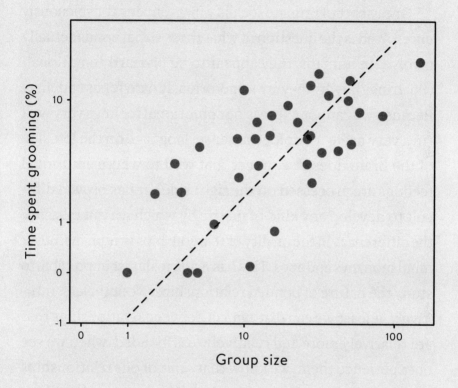

Figure 2.1

Mean percentage of the day devoted to social grooming in Old World monkeys and apes, plotted against mean social group size for each species. Source: Lehmann et al. (2007).

important social partners to the exclusion of casual grooming. Monkey mothers placed under time stress by the feeding demands of their babies also withdraw from casual acquaintances and focus what time they can spare for grooming on their core social partners, the ones that really matter.

One aspect of primate sociality that remains conspicuously unresolved is the question of what these social bonds actually involve. In humans, they appear to be primarily emotional, which may be why they are so notoriously hard for us to define. Because we cannot express our emotional feelings very well (in a very naïve neurological sense, language on the left side of the brain doesn't connect that well to where emotional feelings are processed on the right side), it has proved difficult to develop any kind of metric by which we can measure the differences in the quality of relationships in humans, never mind monkeys and apes. This has made it almost impossible to study the nature of bonded relationships. Nonetheless, intuitively at least we can distinguish between relationships that are relatively more and relatively less 'bonded' when we see or experience them: we know that *some* of our relationships are stronger than others, even though we may not be able to say exactly why.

One rather crude measure that we do have for both monkeys and ourselves is how much time we devote to each of our friends. Our studies indicate that time spent in face-to-face interaction is important in maintaining the quality of a relationship: when we asked people to assess their own relationships on a simple 1–10 scale (1 being 'rather neutral', 10 being 'I love them dearly'), we found that the ratings correlated very strongly with the frequency of interaction.

Indeed, the emotional quality of friendships (but not that of relationships with family members) declines precipitously when friends no longer see each other as much as they have previously done.

It turns out that this finding is consistent with Sternberg's very influential tripartite theory of love, perhaps the most widely accepted social psychological view of relationships. Sternberg defines love in terms of three independent dimensions: intimacy, passion and commitment. Ignoring the passion element as being a peculiarity of romantic relationships, the other two suggest that relationships have two key components: 'being close' and 'feeling close'. These two components seem to correspond quite nicely to the elements of our dual-process model of bonded relationships in primates: 'being close' seems to reflect the grooming element in primate relationships (the need to be in close physical proximity and the emotional attachment that this engenders) and 'feeling close', though undoubtedly a much more nebulous concept, seems to reflect the cognitive component (the sense of being willing to do anything for the partner – including, presumably, coming to their aid, or of having that sense of trust and obligation that is so essential in making social alliances work).

How cognition underpins sociality

Although the cognition that underpins this sense of 'feeling close' is far from clear, one thing that comparative and developmental psychologists agree on is that it involves some form of 'social cognition'. Our best guess as to what

Figure 2.2

Mindreading (mentalizing) involves a reflexive series of mental states about the beliefs/desires/intentions (intentional states) of oneself and others. From left to right, the three individuals are in first, second and third order of intentionality, respectively. ©2014 Arran Dunbar.

this entails is what has become known as 'theory of mind', mindreading or mentalizing. Theory of mind gets its name from the fact that it defines a state in which individuals 'have a theory of mind' – that is, they have an informal theory or belief about what having a mind is like. Practically speaking, this means that they recognize that other individuals have minds like their own. Formally, mentalizing involves the capacity to use words like *believe, suppose, imagine, want, understand, think* and *intend*. Philosophers of mind refer to these kinds of words by the general term *intentionality*, meaning the capacity to take an intentional stance or view.

Intentionality in this sense forms a naturally recursive hierarchy of mental states known as the *orders of intentionality*. All conscious organisms know the contents of their own minds, and being able to do this is referred to as first order intentionality. Formal theory of mind (the capacity to have a belief about someone else's belief) then constitutes second order intentionality: there are two mind states involved, yours and the other person's. After that, there is an infinitely recursive sequence of mind states for the third, fourth, fifth and so on orders of intentionality that can either involve me reflecting on your beliefs about what I believe or me reflecting on your beliefs about someone else's beliefs (Fig. 2.2).

Humans are not born with theory of mind, though they probably soon become self-aware (first order intentionality). Children acquire full theory of mind at about five years of age, and then gradually progress up through the levels until they reach normal adult competence at fifth order intentionality sometime in their teens. At that stage, humans can handle statements like: 'I *believe* that you *suppose* that Peter *wants* Susan to *believe* that Edward *intends* [something or other].' While our studies of normal adult humans suggest that there is some variation around this (most adults vary between fourth and sixth order), the average is consistently fifth order. Only about 20 per cent of the adult population can do better than fifth order.

Brain imaging has revealed that mentalizing is processed in quite specific areas of the brain known as the theory of mind (or ToM) network that involves parts of the prefrontal cortex (there is some disagreement as to exactly which bits are crucial) and two main areas in the temporal lobe (in particular, the frontal pole and the temporo-parietal junction – the area where the temporal and parietal lobes meet) (see Fig. 2.3). We have shown, using both reaction time tests and neuroimaging, that mentalizing tasks are significantly more demanding than equivalent factual memory tasks and that there is a correlation between the amount of neural activity in the core mentalizing network and the level of intentionality at which someone is working. The higher the level of intentionality required by the task, the more neurons have to be recruited to get the answer right. More importantly, we found that people who can work at higher orders of intentionality have larger orbitofrontal regions in the prefrontal cortex (the part just above and behind the eyes). The social brain really is genuinely costly, and, since the volume of neural matter recruited to handle the higher orders of intentionality is proportional to the level of intentionality at which the subject is working, species that need to be able to work at higher orders of intentionality will need bigger brains. It is, of course, significant that it is precisely these

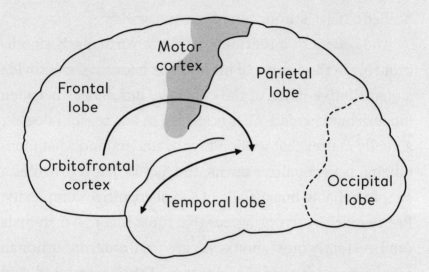

Figure 2.3

The human brain, showing the main lobes. The arrows indicate the mentalizing (or theory of mind) circuit between the orbitofrontal cortex and the temporal lobe. The occipital lobe is almost entirely devoted to vision. The motor areas (shaded) that orchestrate the planning, control and execution of voluntary movements (including facial expressions) lie just in front of the central sulcus (or fissure) that separates the top part of the frontal lobe from the parietal lobe.

frontal parts of the brain that have evolved most recently in anthropoid primates, and are largest in those species that are socially most complex. They are also the last parts of the brain to myelinate (acquire the fatty sheaths that allow neurons to work efficiently), reflecting the fact that a great deal of social learning and neural adaptation is involved in learning the skills needed to navigate our complex social worlds.

The orders of intentionality play a particularly significant role in the story told in this book because they provide a quantitative index of the cognitive differences between modern humans and other primates. In one sense, it doesn't actually matter that we don't really understand what mentalizing is in cognitive terms, the fact is that it provides a simple, reliable linear scale of social cognitive complexity. Pretty much everyone agrees that most if not all mammals (and certainly most monkeys) are first order intentional: monkeys understand the contents of their own mind, and are aware that they have beliefs about the world. There is some experimental evidence to suggest that great apes (specifically, orangs and chimpanzees) can just about cope with second order intentionality (formal theory of mind) – they are certainly not as good as five-year-old children (who pass these tests with ease), but they are about as good as four-year-olds who are just beginning to get the hang of the idea. And at the other end of the scale, our data show that normal human adults are fifth order intentional. If we plot these performances against frontal lobe volume, we get a surprisingly tight linear relationship (Fig. 2.4), suggesting that neural volume in these brain regions is directly related to mentalizing competences in a way that mirrors the relationship

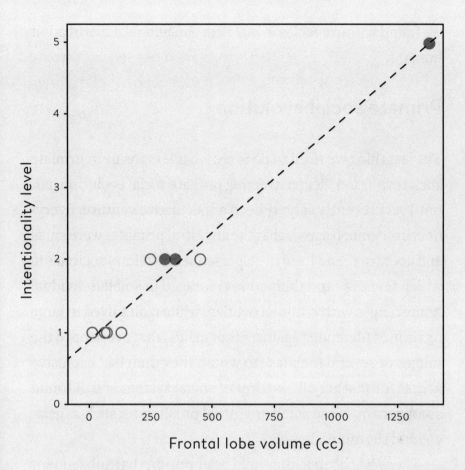

Figure 2.4

Mentalizing competences of Old World monkeys and apes
(indexed as the maximum achievable order of intentionality),
plotted against frontal lobe volume. Monkeys are generally
assumed to achieve only first order intentionality; there is
experimental evidence to suggest that chimpanzees and orangs
can manage second order (just), and that adult humans typically
achieve fifth order.

- ● species for which mentalizing competences have been
  estimated experimentally (left to right: chimpanzees, orang
  utans and humans);

- ○ species for whom mentalizing competences are not known
  but who are assumed to have the same mentalizing competences as other members of their taxonomic family. Note how
  closely clustered known and unknown values are.

Redrawn from Dunbar (2009).

we found within species in our neuroimaging studies of adult humans.

Primate social evolution

The last thing we need to do is see what lessons lie in primate social evolution. Reconstructing primate social evolution has, until very recently, largely been a speculative venture. Everyone pretty much agrees that the ancestral primates were small and nocturnal, and lived in dispersed semi-solitary societies in which females (and their young) foraged in small individual home ranges with minimal overlap, while males lived in large territories (defended against other males) that overlapped the ranges of several females (to whom they then had exclusive access for mating). This form of social system is still found among many of the small nocturnal prosimians, such as galagos and the mouse lemur.

There has been a long-held assumption that subsequent social evolution involved the gradual coalescence of these individuals into permanent groups, with group size largely determining the form of social organization that emerged. One obvious trajectory would be for a male to start associating closely with one of the females in his territory, and so form a monogamous pair. If other females were attracted to join the pair, it would become a harem or single male group (one male with several females), and then if the presence of several females attracted other males to join it would eventually become a large multimale/multifemale group. This is pretty much the standard model of social evolution that you would find in any textbook, and it became known as the socioecological model: a species' social system was simply a consequence of its demographic arrangements, and these in turn were a consequence of its foraging ecology.

However, two recent studies led by Susanne Shultz and Kit Opie suggest that this view is about as wrong as it can be. What made these analyses possible was the development of new, very sophisticated statistical methods that allow us both to estimate ancestral states and to test hypotheses about the historical correlations between social organization and different selection pressures.[2] These analyses suggest that by far the most likely change from the ancestral state of dispersed, individual territories and solitary foraging was directly into a multimale/multifemale form of sociality, and not into monogamy. In other words, animals that had previously ranged alone began to gather together in groups, presumably in response to the increasing threat of predation (mainly associated with the switch from a nocturnal to a diurnal lifestyle). From this initial multimale/multifemale state, there were two possible exits, one to harem forms of sociality and the other into monogamy. There was also a second route into monogamy from harem-based systems. Species never returned to a semi-solitary state once they had formed groups, but did switch back and forth between the harem and multimale states (much as the socioecological model assumed).

The really important finding, however, is that there is no route *out* of monogamy: once a species has opted for this state, it seems that it can never escape from it. In effect, it appears that monogamy is a kind of demographic and cognitive sink, probably because the cognitive demands of monogamous partnerships are so great that once the brain has been rewired to accommodate them, this cannot easily be undone. Monogamy requires the male and female to be very tolerant of each other, but at the same time very *intolerant* of all other members of their own sex. For this reason, monogamous primates always end up as territorial, with each monogamous pair occupying its own exclusive territory. This intolerance of same-sex individuals is quite unusual in mammals outside of monogamous species and makes it very difficult for several individuals of the same sex to live together, especially once they become reproductively active at puberty. Thus, as in other birds and mammals, full-blown obligate monogamy (where every individual is monogamous all the time) appears to be a very special evolutionary state that requires major changes in behaviour and cognition. Once these are in place, there is no going back because the changes are difficult to undo. This will be an important point to remember when we later come to consider the evolution of human pairbonded (or 'monogamous') relationships.

Given that monogamy has a very specific evolutionary history, the question of why monogamy should have evolved as a social and mating system needs to be considered. Over the years three different explanations have been offered for monogamy in mammals. These are: (1) the need for biparental care (two parents are needed to raise big-brained offspring), (2) mate-guarding by males (when females are so widely dispersed that a male cannot defend more than one at a time, he sticks with one female in order to ensure that he at least fertilizes her when she becomes receptive, and at the same time prevents other males from getting a look-in), and (3) infanticide risk (a female locks on to a male in order to use the male as a 'hired gun' or bodyguard to defend her against other males that might harass her and/or kill her offspring). Infanticide risk has long been recognized as a serious problem for primates: because primates' large brains slow down their reproductive rate (because it takes a long time to grow a large brain), a male who takes over a female from another male may have to wait a year or more before being able to sire his own offspring. However, if the male kills the female's current infant, she will return to breeding condition immediately (the same is true of women who lose an unweaned infant, by the way)[3] and the male can get started on his reproductive career straight away. Because of this, the argument runs, there is intense selection pressure for infanticidal behaviour in male primates, and infanticide is surprisingly common in primates (despite valiant attempts to claim that it is rare); and this in turn creates an equally intense selection pressure for counter-strategies that mitigate or reduce this risk.

The second study examined whether the switch into monogamy occurred in the various primate lineages before or after the change in behavioural indices that represented each of the three hypotheses. Although there is evidence of co-evolution between monogamy and females living in large, discrete home ranges (the two traits change together through time), female range size doesn't increase *before* monogamy evolves, suggesting that the crucial premise on which the mate-guarding hypothesis is based (females become too widely dispersed for males to monopolize several of them at the same time) does not hold. In fact, for no monogamous species of primate is it true that males couldn't defend more than one female: the males of these species invariably range widely enough to allow them to defend the territories of several females *if they really wanted to do so*. Mate-guarding due to females being in dispersed ranges is certainly the explanation for the evolution of monogamy in most mammals, but not, it seems, in primates. Similarly, there is also evidence of co-evolution between monogamy and biparental care, but monogamy evolved just as often in the presence of biparental care as it did in its absence, indicating that biparental care is more likely to be a *consequence* of monogamy rather than its cause: once you have monogamy in place, it pays males to contribute to parental care because they rear more offspring by doing so. But the male's ability to offer parental care seems not to be sufficiently advantageous on its own to drive the evolution of monogamy, at least in primates. Finally, there is strong co-evolution between monogamy and two separate indices of infanticide risk, but the switch from polygyny to monogamy never occurs in the absence of high infanticide risk. In primates at least, infanticide seems to have been the crucial factor driving the evolution of monogamous mating systems.

Pairbonded monogamy in which the male provides protection for the female is not, however, the only solution to risk of infanticide. Sandy Harcourt and Jonathan Greenberg showed that gorillas have gone for a hired gun solution as a result of high infanticide risk, but have combined this with harem-based polygyny: several females attach themselves to the same male protector. In this case, the massive size dimorphism[4] between males and females in this species may well have made the largest and most powerful males particularly attractive as protectors compared to the smaller males, at the same time setting up a feedback mechanism that drove males to becoming ever larger. In conventionally monogamous primates like the gibbons and smaller New World monkeys, by contrast, males and females are the same size (indeed, females may even be slightly larger than the males). The social arrangements of gorilla groups are quite different to those in the more social monkeys, where the females invariably form grooming-based coalitions. Gorillas have a star-shaped social structure, with the male at the centre and the females forming the arms of the star: each female grooms with the male, but females rarely groom with each other. While the Harcourt–Greenberg model provides a convincing explanation for harem-formation in gorillas, it also shows rather clearly that the infanticide risk is not high enough in chimpanzees to make the hired gun strategy worthwhile for chimpanzee females.

We will need to bear these general principles in mind when we come to examine hominin social evolution, since we must necessarily assume that hominins, like all primates, will be subject to these same social and reproductive selection forces. The fundamental point is that as social group size increases, females will face increasing stresses and males will be forced into competition with each other. If they cannot find solutions that defuse these constraints, they will not be able to occupy new kinds of habitats or evolve into new species with larger brains. The inevitable result will be extinction when climate change results in dramatic reduction in its favoured habitats. Since we know that they did survive, we can be certain that they must have solved these problems somehow. This brings us to our two fundamental principles that underpin the story of human evolution: brain size and time budgets. Brain size determines social group size as a response to environmental conditions, and group size and environmental conditions between them impose demands on time budgets that must be satisfied if a new evolutionary step change is to be possible. In the next chapter, I set out my stall in more detail so as to establish exactly what these can tell us about each of the five phases of hominin evolution.

The Essential Framework

As I suggested in Chapter 1, the framework that I will be using to explore human evolution consists of two key elements, the social brain hypothesis and the time budget models. Between them, these provide a scaffold on which we can pin each of the main species of hominin as they appear in the fossil record in a way that allows us to ask how they coped with their respective circumstances. At the same time, they also provide the criterion that will allow us to determine the novel components of their behaviour and cognition: time budgets have to be able to accommodate the demands that the changes in brain size show us happened.

The social brain hypothesis

There is a general consensus that the prime mover in primate brain evolution (and perhaps even that of all mammals and birds) is the evolution of more complex forms of sociality. Although other aspects of behaviour (notably ecological inventiveness) also correlate with brain size, these turn out to be consequences of having a large brain rather than its evolutionary cause. In most mammals and birds, the social brain hypothesis appears as a relationship between brain size

and the mating system, with monogamously pairbonded species having significantly larger brains than species that mate
polygamously or promiscuously, and especially so if monogamy involves lifelong pairbonds. We think this is probably
because longlasting pairbonds are cognitively much more
demanding than the more casual relationships of species that
mate promiscuously. Pairbonded individuals have to be able
to factor their partner's interests into their decisions on what
to do; they need to be able to negotiate between their respective needs to find some appropriate compromise – in effect,
a primitive form of mentalizing, even if it is not full-blown
theory of mind. Incidentally, the social brain hypothesis also
seems to hold for insects: among wasps and sweat bees, social
species (such as those where several queens share a nest) have
larger brains – or to be more specific, mushroom bodies (the
bit of the brain that manages higher cognitive and especially
social functions) – than solitary species. Moreover, within
species more social individuals (such as queens) have larger
brains than less social ones (such as workers). Something similar may even apply in fishes.

In primates, and perhaps a small number of other
mammal families (notably elephants and the horse family),
the social brain effect is converted into a quantitative relationship between a species' brain size and the average size
of its social groups. This is because all these species have
bonded social relationships. These are what we might think
of as friendships – intense, emotionally close relationships
that don't involve sex and reproduction. These contrast
starkly with the rather loose here-today-and-gone-tomorrow
relationships we find in most herding mammals such as deer

and antelope. Primate 'friendships' are similar to pairbonded relationships and, like reproductive pairbonds in other mammals and birds, require more brain power to manage; the number of such relationships that an individual can manage is then a simple function of how big its brain is.

The social brain relationship manifests itself in primates as a cognitive limit on social group size that gives rise to a correlation between social group size and brain size (Fig. 3.1). The social brain hypothesis has been given a significant boost by a recent spate of neuroimaging studies demonstrating that the absolute volume of key brain regions correlates with social network size at the individual level (including the number of Facebook friends), both in humans and in monkeys. The important point here is that the neuroimaging studies compared individuals within the same species, thus demonstrating that the social brain hypothesis applies not only at the species level but also within species at the individual level.

I will use the term brain size loosely throughout most of this book: in fact, the relationship is with neocortex volume, with evidence from both the comparative and neuroimaging studies to suggest that it is frontal lobe volume in particular that is crucial. More importantly, the relationship is really one between *behavioural complexity* and brain (or neocortex) size, with group size itself being an emergent property: the number of relationships an individual can maintain depends on the complexity of its social behaviour, which is in turn dependent on its cognitive abilities (and hence brain size). We see this quite clearly in primates: several behavioural indices correlate with neocortex volume, including grooming

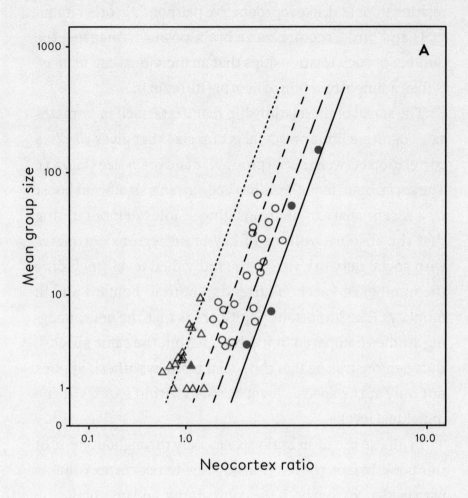

Figure 3.1

The social brain relationship. Mean species social group size
plotted against (a) neocortex ratio and (b) frontal lobe ratio.
In both cases, the ratio is relative to the subcortical brain.

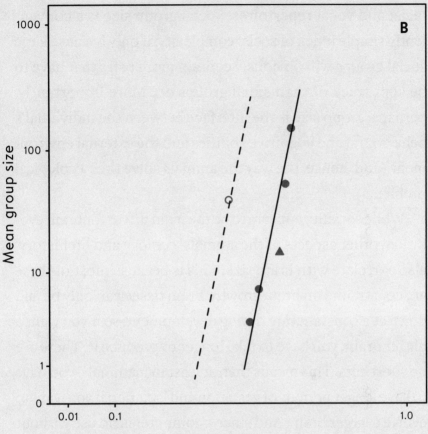

- Apes: filled circles and solid line;
- monkeys: open circles and dashed lines (reflecting different socio-cognitive grades within the monkeys);
- prosimians: triangles and dotted line.
- In (b), the orang utan is represented by the filled triangle; the monkey is the macaque.

Reproduced from Dunbar (2010).

clique size, the use of coalitions, the use of deception, the kinds of mating strategies males use, and the complexity of facial and vocal repertoires. Social group size is a conveniently simple index of social complexity, if only because large social groups will obviously contain more pairs that have to be kept track of than small groups do. More importantly, perhaps, group size is the interface between the individual's behavioural and cognitive abilities and the external environment (and, hence, the way the animals solve their ecological problems).

While sociality appears to be the main driver for brain evolution, other aspects of the animals' ecology and lifehistory[1] also correlate with brain size. This is because most of these are constraints on brain growth. Brain tissue can only be laid down at a constant rate during development, so if you want a bigger brain, you have to take longer to produce it. There are no short cuts. This means that, at least in mammals, you have to have longer periods of gestation and lactation if you want to evolve a larger brain. And since a computer is no use without software, you also have to have a longer period of socialization (essentially the time between weaning and the start of reproduction) in order to allow the brain to fine tune its ability to deal with all the subtleties of the dynamic, constantly shifting social world. Neuroimaging studies of humans suggest that it takes a surprisingly long time (around 20–25 years) for the brain to figure out how to handle the complexities of our social world.

Brains are very expensive both to grow and to maintain. An adult human brain consumes about 20 per cent of our total daily energy intake, but represents only about 2 per cent of our body weight – so it consumes about ten times more energy than we would expect for its mass (and that is just to keep it alive and doesn't include the costs when it is busy sorting out what we should do). Thus, having efficient foraging strategies that make it possible to acquire enough extra food to fuel the brain becomes an increasingly important constraint on a species' ability to evolve a larger brain. Some diets (notably leaves) may be too time-consuming or poor in quality to allow a species to evolve a large brain without changing its diet. We can think of this as a kind of grey ceiling (most of this concerns the grey matter of the brain) that sets a limit on what a species can do cognitively, and hence socially, with a particular diet.

Monkeys and apes (the anthropoid primates) seem to differ from all other mammals (including prosimian primates) in having evolved entirely new brain regions in the prefrontal cortex. In their seminal book on primate brain evolution, the neuropsychologists Dick Passingham and Steven Wise argue that these prefrontal areas are associated with sophisticated new cognitive capacities that are unique to anthropoid primates, including the ability to make causal leaps of inference (one-trial learning) and to project and compare behavioural outcomes into the future. Our neuro-imaging studies add an important additional dimension: we showed that there is a specific causal sequence in which orbitofrontal cortex volume determines mentalizing competences, and mentalizing competences determine the size of the social network. This being so, the scaled up differences in mentalizing ability between monkeys, apes and humans that we see in Fig. 2.4 are likely to have dramatic implications for what these species (and the fossil hominins that are likely to fall on the line between them) will be able to do both socially and culturally, as we shall see in Chapter 8.

The prefrontal cortex also appears to be associated with the ability to inhibit impulsive responses (technically, to inhibit prepotent responses) so as to allow the animal to postpone a reward until later rather than grab what it can now – something that humans are, relatively speaking, particularly good at. Being able to postpone a reward in this way is crucial to living in large, bonded social groups because this is only possible if each of us is willing to forgo some of our immediate selfish desires in order to allow everyone else to have a fair share of the cake. Species with larger frontal lobes are better able to control their impulses – to suppress the red mist that leads to an angry response when you are affronted by someone else's behaviour.

Humans and the social brain

The bottom line for our story is that the social brain hypothesis provides us with a precise equation for predicting social group size from brain size. So what does the social brain tell us about modern humans – the end-point that our evolutionary story will aim at?

It has become clear that the data in Fig. 3.1 in fact form a number of parallel grades. This is most obvious for the apes, who sit out on one side of the graph. In effect, these grades reflect increasing levels of socio-cognitive complexity as one moves from left to right. Because humans belong to the ape family, we need to use the equation for the ape grade rather than the one for primates as a whole when estimating human group sizes. Interpolating the neocortex ratio for modern humans into the ape equation gives a predicted group size of approximately 150. Do humans really live in such small communities?

It seems that they do. One answer comes from looking at census data for hunter-gatherer societies. These societies reflect the kind of small-scale social organization that we have lived in for most of our evolutionary history. They are thus an especially appropriate place to test this prediction, given that human brain size hasn't changed much in the last 200,000 years. In effect, this is to look down from above on the distribution of people in space and describe the spatial structure of the community. Much as most primates do, hunter-gatherers live in a multi-level form of social organization composed of several inclusive layers. In hunter-gatherers, these consist of families, camp groups (or bands), communities (or clans), endogamous communities (or mega-bands), and ethnolinguistic units (or tribes). Fig. 3.2 shows the distribution of sizes for the different layers of hunter-gatherer society above the family level (which is obviously too small to be relevant for the social brain story). It is glaringly obvious that only one of these grouping levels has a typical size of 150, namely the layer I identify as the community. In most hunter-gatherer societies, this layer is associated with a group of people who own rights of access to territory or to special resources like permanent waterholes or sacred sites; they also meet together for events like rites of passage, usually on an annual basis.

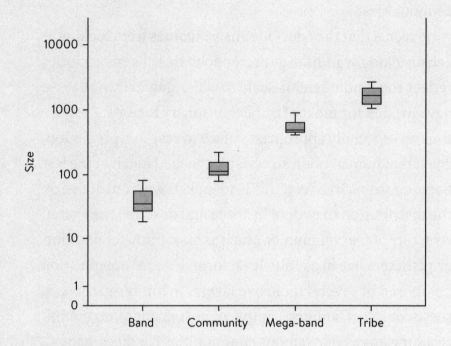

Figure 3.2

Median (with 50 and 95 per cent range sizes) of the main grouping layers in small scale hunter-gatherer and horticultural societies. Source: Dunbar (1993).

Table 3.1 illustrates how common groupings of about 150 are in human social organization. Examples include historical village sizes, parish sizes for various Christian denominations, and business and military organizations. In modern armies, for example, the smallest unit that can act on its own is the company, and this has an average size of almost exactly 150 (with a range of about 120–180).

An alternative approach is to get people to list out their entire social networks. In effect, this is to look at the world bottom-up from the individual's viewpoint. We have done this several times, in some cases by asking people to list all the recipients of their Christmas cards, in others by asking them to work through their address book and list all the people that are meaningful to them. One study of Christmas card lists produced an average of 154 – comfortably close to the figure of 150. Another large sample of women asked to list all their network members produced a figure of around 150 as the upper limit. Other researchers have analysed the size of Twitter communities (those who exchange posts with each other as followers of a Twitter account, based on a sample of 1.7 million people) and email communities (those who exchange messages, in this case based on a sample of 10 million people), and also found that the size of interacting communities was in each case typically between 100 and 200 individuals. A recent survey of 1 million Facebook accounts revealed that the number of friends listed is very skewed: a few people do indeed list as many as 5,000 friends, but the number who list more than 500 is tiny, and the vast majority have only around 150–250 friends (and even then, they are not always sure who all of these actually are!).

	TYPICAL SIZE
Neolithic villages (Middle East, 6500–5500 BC)	150–200
Maniple ('double century') (Roman army: 350–100 BC)	120–130
Domesday Book (AD 1085): average county village size	150
C18th English villages (mean of county means)	160
Tribal societies (mean and range of communities; N = 9)	148 (90–222)
Hunter-gatherer societies (mean clan size; N = 213)	165
Hutterite farming communities (Canada) (mean; N = 51)	107
'Nebraska' Amish parishes (mean; N = 8)	113

	TYPICAL SIZE
Anglican church congregations (recommended ideal size)	200
East Tennessee (USA) rural mountain community	197
Social network size (mean; N = 2 'small world' experiments)	134
Gore-Tex Inc: factory unit size	150
Company (mean and range Second World War armies: N = 10)	180 (124–223)
Christmas card distribution lists (mean total recipients: N = 43)	154
Research specialities (sciences and humanities) (mode, N = 13)	100–200

Table 3.1

Community sizes in a selection of historical and contemporary human populations

Given that the social brain hypothesis predicts a 'natural' community size of about 150 for modern humans, and that there is considerable casual as well as systematic evidence for this, the final thing we need to do before moving on is to determine what the social brain equation predicts for the broad sweep of fossil hominins. This will provide us with the crucial framework within which to evaluate hominin behaviour at each stage in human evolution. We will do this using cranial volumes corrected for the fact that the brain doesn't occupy the whole cranium. A note of caution needs to be sounded here because brain regions can sometimes evolve at different rates depending on the ecological pressures acting on a species. The gorilla and orang utan provide examples of this: both have smaller neocortices than you would predict given their total brain volume – mainly because they have much bigger cerebella,[2] probably in order to allow them to manage their very large bodies in a three-dimensional environment in trees. Using cranial volumes would result in the size of their neocortex (or worse still, the frontal lobe) being overestimated in these species, and so cause social group size to be overestimated too. The cerebellum is involved in, among other things, managing balance and is therefore relatively larger in bipedal humans than quadrupedal primates since maintaining balance during bipedal walking is especially challenging. This means that there is likely to be some risk of overestimation when we use cranial volume to estimate group size in all fossil hominin species once bipedalism became the norm. However, let's ignore this for the moment and see what cranial volume alone tells us about fossil hominin community sizes.

When we use cranial volume to estimate group (or community) size for individual fossil hominin populations using the great ape social brain equation, we get the graph shown in Fig. 3.3. Of course, these data just map changes in cranial volume (see Fig. 1.3), but the important thing for us is that they provide a specific value for community size for each hominin population, as well as the degree of variability in this. What really matters for our project, however, is that each species has to be able to maintain its typical community size as a coherent social entity, as well as cope with the implications this has for time budgets. Palaeoanthropologists have been known to comment that the data like those in Fig. 3.3 don't really mean anything because fossil hominins could have behaved very differently to modern humans and so had quite different kinds of brains. In fact, this isn't an especially sensible suggestion, given that these values are anchored at either end with precisely known group sizes for chimpanzees and modern humans. Unless we are prepared to claim that chimpanzees and modern humans (and all other living primates) abide by the social brain hypothesis but for a few million years in between fossil hominins mysteriously had a completely different kind of brain or social style to all other primates (including modern humans), then the only issue is how the various hominins are distributed between these two end anchor points. The most parsimonious assumption is that their group sizes simply follow the changes in brain size – unless we have principled reasons for believing otherwise. And so far, no one has suggested any principled reasons.

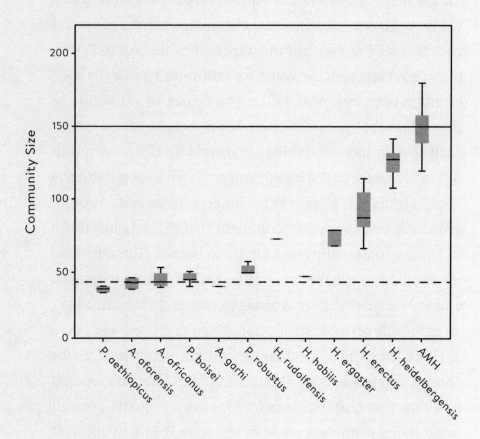

Figure 3.3

Median (with 50 and 95 per cent range) community sizes for the main hominin species, predicted by converting cranial volumes (from Fig. 1.3) to neocortex ratio, and then interpolating these into the ape equation from Fig. 3.1. Solid horizontal line denotes community size in modern humans (150) and the dashed line the average community size for chimpanzees. Note that, for reasons to be discussed in Chapter 6, Neanderthals have not been included in this graph.

When a community isn't quite a community

Let me pause for a moment to deal with another issue that has been a perennial thorn in the side of the human social brain story. You might by now suppose that the idea that humans have a natural grouping size of around 150 is signed, sealed and delivered. Unfortunately, nothing could be further from the truth. Despite the fact that the social brain relationship makes a very specific prediction for the natural grouping size of humans and that there is considerable evidence to confirm this, mere evidence has not prevented debate as to what the 'true' community size really is for modern humans.

Archaeologists typically point to the fact that the only kinds of communities they can see in the archaeological record are overnight camps, and there has been a long tradition in social anthropology of identifying these camp groups (or bands) as the fundamental social unit for contemporary hunter-gatherers. These camp groups typically vary from 30 to 50 in size, depending on the local environment and the economy of the people concerned. At the other extreme, some social scientists have argued that natural human groups are much larger than 150, pointing to the inherent coherence of tribes in traditional societies or to the large size of personal 'networks' that some people claim to have on internet social networking sites. This latter view typically argues for a natural community size that is much greater than 200, perhaps as large as 500–1,000 individuals.

This diversity of views is in itself interesting, but it creates a problem for us. Unless we know exactly what the natural human group size is, we won't know what the hominin trajectory was supposedly aiming at. Indeed, if the archaeologists are right and the band of 50 *is* the basic human community, then there has been no real social evolution since we parted company from the chimpanzees (who have an average community size of 55). And that means there is nothing to explain – in which case, I needn't have bothered writing this book! Conversely, if the sociologists are right and natural human group sizes are very large (or there really isn't such a thing as a natural group size), then the social brain story with its emphasis on groupings of 150 may be radically misdirecting us.

In fact, the resolution of this apparent dilemma is very straightforward. The ethnographic evidence tells us that hunter-gatherer camp groups (bands) are not, in fact, the fundamental units of human social organization, because they are actually quite unstable: their membership changes over time on a scale of months as individuals or families decide to join or leave. The important point is that when a family does join another camp group, it invariably joins one whose members belong to the same 150-member community (bonded community or clan); they rarely join camps from a completely different community (unless, of course, they are under extreme ecological duress or have been banished). Equally, while it is true that most of us, including contemporary hunter-gatherers, know more than 150 people, there seems to be a radical distinction between those within the 150 layer and those outside it. We typically think of the people in the outer layers beyond 150 as acquaintances: they are not friends and relations in the meaningful sense of these terms, but rather people we know casually by sight. Our research over the last decade points to a striking division at the outer edge of the 150 layer: those within are people we know as individuals, based on relationships that have historical depth and involve trust, obligation and reciprocity; they are the people that we don't really think twice about helping when they ask. In contrast, our relationships with those in the layers beyond 150 are more casual, unreciprocated and often lack history; we are demonstrably less generous to people in this outer layer. Yes, we can always add more 'friends' on to our Facebook page, but we are not creating more *friendships* in any real sense: all we are doing is extending our Facebook network out into the acquaintances layer (the 500 layer) that naturally exists in our face-to-face world.

In short, the alternative grouping layers that others have identified certainly exist. In fact, they are part of a hierarchically inclusive sequence of grouping layers as Fig. 3.2 (see also Fig. 3.4). The social brain relationship specifically identifies the 150 layer as the grouping for modern humans that is equivalent to the natural groups of other monkeys and apes. We can always identify other grouping layers as being important, as different groups of researchers have done, but these layers are not actually all the same kind of thing: they represent very different kinds and qualities of social relationship, and, as we shall see later, they serve very different functions for us. The 150 layer is, as we shall see in Chapter 8, important for another, explicitly social reason: it demarcates the limits within which we recognize kinship (no human culture has formal kinship terms for anyone that inhabits the layers beyond 150). Since this is the layer that corresponds to primate social groupings in the social brain relationship, it is the size of this layer that we need to worry about.

Structural complexity in primate social systems

We have tended to think of social groups as though they were socially homogeneous (everyone is friends with everyone else). While this is broadly true of most species, large primate groups are in fact highly substructured, rather in the way that human communities are (Fig. 3.2). It is probably this structuring (created by the fact that individuals focus their attention increasingly on their core allies as grooming partners) that allows primates to live in larger groups because, as we saw in Chapter 2, it is these coalitions that buffer them against the costs of living in large groups. At the same time, this structuring creates social complexity, thereby imposing the cognitive demand that is so characteristic of the anthropoid primates.

It is obvious from Fig. 3.2 that hunter-gatherer communities are one layer in a hierarchically nested series of levels of social organization. In this respect, hunter-gatherer societies are typical of all human societies (we all live in nested social layers of this kind) and, as it turns out, most monkeys and apes. To explore this further, we analysed the distribution of grouping sizes in the ethnographic dataset shown in Fig. 3.2 using the mathematics of fractals to search for repeated patterns in the data. This revealed that these social layers have a very distinct scaling signature: each layer is three times the size of the layer inside it. For our hunter-gatherer dataset, this yields a series of layers at 50, 150, 500 and 1,500: three camp groups of around 50 individuals make up a bonded community (or clan), three bonded communities make up an endogamous community (or mega-band) and three endogamous communities make up an ethnolinguistic unit (or tribe). Using the frequencies of contact in our Christmas card dataset (which focuses on relationships within the 150 circle) yielded exactly the same pattern: a scaling ratio of almost exactly three, identifying a set of inner layers at 5, 15, 50 and 150. So between them, these two datasets suggest a natural sequence of grouping layers running from an innermost core of 5 to an outermost grouping of 1,500 in a very distinct pattern: roughly 5–15–50–150–500–1,500. Gratifyingly, another study subsequently showed the same pattern in a different hunter-gatherer dataset.[3] More interestingly, we also found the same scaling ratio in the hierarchical structure of the social systems of other mammal species that live in complex societies (chimpanzees, baboons, elephants and orcas), suggesting that this pattern may be widely characteristic of mammals that have complex social systems.

Fig. 3.4 shows what this looks like when seen from your personal perspective: you sit at the very centre, surrounded by a series of layers with each layer progressively including more people. In humans, we might think of these layers as intimate friends (the innermost 5 layer), best friends (the 15 layer), good friends (the 50 layer), friends (the 150 layer), acquaintances (the 500 layer) and people whose faces we can put names to (the 1,500 layer). In our contemporary societies, the four innermost layers (out to 150) consist of roughly half family and half friends, and the layers beyond

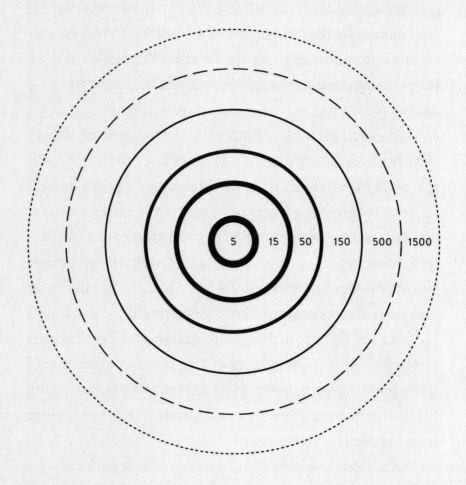

Figure 3.4

The circles of acquaintanceship. Our social networks consist of a series of hierarchically inclusive layers, with each layer being three times bigger than the layer immediately inside it. Beyond the layer of 150 that makes up our personal social network lie at least two further layers, one at 500 (acquaintances) and a second at 1,500 (the number of faces we can put names to).

150 consist purely of people we know casually or by sight: this layer rarely contains family. One obvious hypothesis is that hominin social evolution consisted of progressively extending out beyond the core chimpanzee community of 50 to add, successively, the 150, 500 and 1,500 layers. We still have no idea why the scaling factor should be almost exactly three, however – even though we find exactly the same scaling ratio in primates, elephants and orcas (killer whales).

It seems likely that the difference between species with more and less complex social systems lies in the number of layers they have rather than the sizes of the layers themselves. Humans, for example, have six layers, while chimpanzees and baboons have only three, and the decidedly less smart colobus monkeys have only one (or at most two) layers. The capacity to maintain these multi-level social systems may well depend on a species having sufficiently developed social cognition to be able to manage several grouping levels at the same time, and hence a large enough brain to support the mentalizing capacities required to do this (Fig. 2.4). Some indirect evidence for this comes from comparative analyses: Bogusław Pawłowski and I found that, across primates, species with bigger neocortices were more likely to integrate two separate social strategies (individual dominance vs social alliances) than species with smaller neocortices. In a very nice set of field experiments, Thore Bergman and Jacinta Beehner showed that baboons can integrate knowledge about an individual's kinship relationships with knowledge about their dominance ranks and keep the two dimensions separate in their minds. Less intellectually well-endowed species like colobus monkeys would not be able to do this, and as a result their groups are structured around the simpler principle of dominance rank, and this probably explains why their groups are smaller.

This structuring really reflects the extent to which an individual interacts with (grooms, in the case of monkeys and apes) the other members of its group. In our case, something like 40 per cent of our total social effort – our social capital, if you will – is devoted to the five people in our innermost circle of relationships, our five most intimate friends and family, and about 60 per cent is devoted to the 15 people in the two innermost layers, with the remaining 40 per cent being divided between the remaining 135 people in the outer two layers (Fig. 3.5). These frequencies of interaction correspond rather closely to levels of emotional closeness, as Sam Roberts and I were able to show. Indeed, we found, in an 18-month study of relationship change, that when the frequency of interaction between friends declined, it resulted in a corresponding decline in perceived emotional closeness. Intriguingly, it turns out, as a follow-up analysis of this same dataset by Jari Saramäki showed, that we each have a very distinctive signature, rather like a fingerprint, in how we distribute our social capital (time and emotional effort) across the individuals in our extended social networks – so much so that this pattern is preserved even when there is considerable turnover in network membership (as happens when someone moves away from their home town).

As with the monkeys and apes, we concentrate our social capital where it is most important – the core members of our close alliance that will provide us with emotional and other kinds of support. But at the same time we retain links with

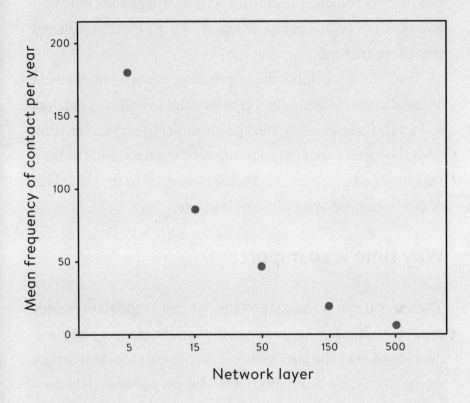

Figure 3.5

Mean frequency per year with which we contact individual members of the different layers of our social network, based on 251 women's complete social networks from the UK and Belgium. Redrawn from Sutcliffe et al. (2012); data from Roberts et al. (2007)

a number of others who provide rather more diffuse forms of support. In the sociological literature, these are often referred to as strong and weak ties (to use the terms coined by the American sociologist Mark Granovetter). Notice that the sociological literature only distinguishes between two types of relationship, whereas Fig. 3.4 clearly suggests that there are four.

These data reinforce the importance of social interaction as the glue that binds human relationships every bit as much as it does those in monkey and ape groups (Fig. 2.1) and remind us of the significance of grooming time as an issue in the time budgets of all primates. This brings us neatly to the other half of our framework, namely time budgets.

Why time is so important

The second major foundation for our task is the time budget models that Julia Lehmann, Mandy Korstjens and I have developed over the last decade or so. These models will play a crucial role in the chapters that follow because they provide us with a way of determining how well fossil hominins coped with their environments. In combination with the social brain relationship and the archaeological data, they will be the fulcrum that helps us understand hominin social evolution.

The time budget models are, conceptually, very simple. They begin from the observation that an animal can survive in a given habitat only if it meets its energy and nutrient requirements and ensures the cohesion of its social group. Nutrient needs are met by allocating time to foraging (which includes both feeding and travel), and social cohesion is satisfied by allocating time to whatever activities enable this – in the case of primates, social grooming. This leaves just one other major activity category to worry about, namely resting time. This is not rest in the sense of having nothing better to do, but time that the animal is forced to spend inactive either to avoid heat overload during the middle of the day or – a particular problem for species with leaf-based diets (folivores) – to allow digestion to take place.[4]

We used data from studies of wild monkeys and apes to develop equations that specify how much time each genus has to devote to feeding, travel and rest as a function of its diet, the local climate and the size of the social group, at a given location. All we need to do is to figure out how much time is left over in the normal waking day (12 hours for tropical monkeys and apes), and then determine how big a group can be bonded in that amount of time (given the relationship between grooming time and group size in Fig. 2.1). The bottom line is that if a species can only manage to bond groups of three or four animals in a particular location, it is unlikely to be able to survive there – not least because it won't be able to cope with the level of predation risk it faces.

Fig. 3.6 illustrates the basic idea of the time budget models and how they work. The starting point is the climate at a particular location, which directly or indirectly (via its effect on vegetation) determines the core time budget elements, and from this we can figure out the maximum group size that can exist there. In all, models have been developed for six African and one South American monkey genera,[5] and all four ape genera (gibbons, orangs, gorillas and chimpanzees).

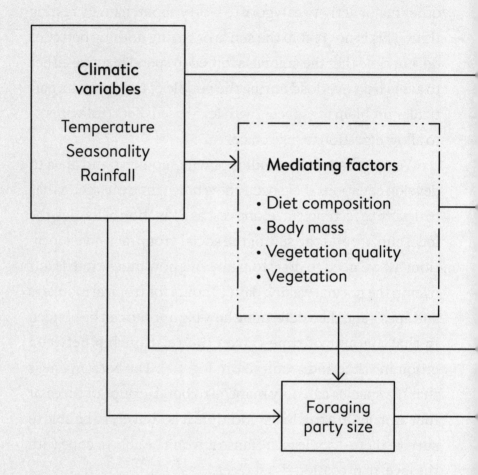

Figure 3.6

Basic structure of time budget models. Climate and the species'
physiological adaptations determine core time budget allocations for
feeding, moving and enforced resting time (the time an animal has to
rest to cope with high temperatures or with digestion). The balance
in time left over (free rest) can be allocated to social time, and it sets
the maximum group size the animals can maintain in that habitat.

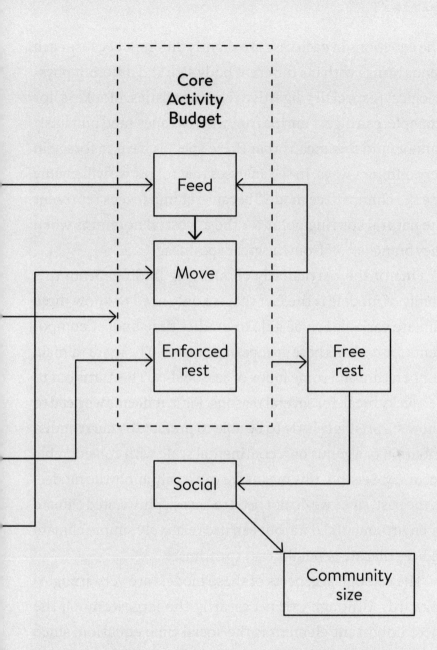

For primates, foraging party size is the main defence against predators; for most species foraging party size and group size are one and
the same, but in fission–fusion social systems (as in chimpanzees)
communities can fission into smaller foraging parties. In great apes,
foraging party size dramatically influences moving time.
After Dunbar et al. (2009).

The equations in each case are slightly different because each genus brings with it a different body size and different physiological (especially digestive) characteristics. Monkeys, for example, can digest unripe fruit whereas apes (and humans!) cannot, and this means that these species have to forage in very different ways. In the analyses that follow we will mainly use the chimpanzee model[6] because chimpanzees represent the natural starting point for the ancestral hominins when they branched off from the great ape lineage.

One of the particularly comforting findings from this family of models is the fact that we only need to know three climate variables to be able to predict time budget components for each of these groups of primates. These are rainfall, temperature and some index of seasonality. This turns out to be a lucky break for several reasons. First, it means we need to know surprisingly little to be able to predict the current distribution of a genus on a continental scale with considerable accuracy. Second, this means we can easily apply the models to the past, since we do not need to have sophisticated climate or environmental data, but can use relatively simple climate indices that are generally easy to estimate.

The ecological aspects of these models are very straightforward. Although not necessarily the largest, much the most important element is the social time equation, since this directly determines the size of social group that a species can maintain. Primates bond their social groups by grooming each other. As a result, social grooming increases linearly with group size across the primates (Fig. 2.1). The fact that the relationship between social time and group size is more or less linear provides us with a simple rubric for determining how much time animals *ought* to allocate to grooming if they want to live in groups of a certain size. If group size starts to exceed the time the animals have available for grooming, given the other demands on their time, then it means that they will be unable to groom everyone they need to groom frequently enough to keep the bond between them at the right strength. As a result, the group will fragment, and eventually break up into two or more smaller groups that go their own way. Notice, by the way, that the time budget model does not specify the size of group a species lives in in a particular habitat; it specifies the *maximum* group size that it can cope with in that habitat. It is not obliged to live in groups of that size (given the costs of doing so: see Chapter 2), but it definitely cannot live in groups that are larger.

Despite the fact that these models are based on only a handful of climate variables, they turn out to be extremely good at predicting where species can and cannot live. We have tested the models by using them to predict the known biogeographic distribution of the various species on a continent-wide scale. We do this by dividing the continent, say Africa, into a matrix of small squares 1° latitude by 1° longitude, determining the relevant climate profile for that square (usually from large-scale climate models), and then using this with the equations from the appropriate model to predict group sizes in the squares and, finally, comparing the resulting distribution with where the species actually lives. The fit is astonishingly good; in fact, it is usually slightly better than the best conventional biogeographic models from conservation biology (probably because the time budget models provide a more nuanced quantitative fit whereas biogeographic models only provide a dichotomous present/absent prediction).

There are a few important findings from the ape models that are crucial to this book, so let me spell them out. First, the key constraint on great ape biogeographic distribution is travel (or moving) time (Fig. 3.7). This turns out to be a consequence of two factors that affect travel time demand in their case: one is the richness of the habitat (indexed by rainfall in the figure) and the other is the size of the foraging party – in part because of competition between group members and in part because, being only able to eat ripe fruit, they exhaust their food patches more quickly. Travel time very quickly becomes prohibitive as the apes move away from wet equatorial forest zones, especially if the community tries to forage as a single group (Fig. 3.7). It is for this reason that apes are restricted to a narrow belt round the equator. In contrast, baboons (genus *Papio*) are limited by the time costs of feeding (i.e. the amount of food they can eat per minute), and are relatively immune to travel time costs (in part, a consequence of their ability to cope with unripe fruit). As a result, the distribution of baboon populations is considerably wider than that of apes: they can live more or less anywhere in sub-Saharan Africa provided it is not desert or dense forest. Given that early hominins lived in a much wider range of habitats than modern apes, it immediately raises the question of how they coped.

Second, chimpanzees are only with us today because they can exploit a form of fission–fusion sociality to offset the high costs of travel that they face. Fission–fusion social

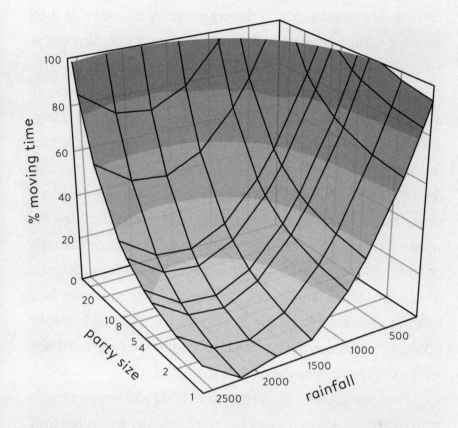

Figure 3.7

The percentage of the day that great apes would have to spend moving from one feeding site to another. The graph plots the amount of time the chimpanzee model predicts that a group would have to devote to moving in a given habitat as a function of the number of animals that forage together (party size) and habitat quality (here indexed as average annual rainfall (mm), where wetter habitats are usually richer). Moving time increases exponentially as party size increases and the habitat gets drier (less forested and more savannah-like). If the typical chimpanzee community of 50–80 foraged as a single group, they would be obliged to spend more than 100 per cent of their day just travelling. In fact, few species of monkeys and apes devote more than 20 per cent of their day to moving. In order to bring their travel time allocation down within reasonable limits, chimpanzees have to split their communities into a number of small foraging parties (typically just 3–5 individuals). Reproduced from Lehmann et al. (2007).

systems are ones in which the community can split up into smaller foraging parties during the day. This allows the animals to reduce the time demand for travel drastically (Fig. 3.7). Were chimpanzee communities forced to forage as a single large group, they would run out of time almost everywhere that they currently live. In many of the drier habitats occupied by baboons they would have to devote most of the day to travel with little time available for feeding or resting, never mind social interaction. Indeed, even in the habitats where they now live, chimpanzees would be able to manage communities of only 10–15 animals at most, and chimpanzee communities this small are not viable. Since chimpanzees face this problem even in relatively rich forest habitats, hominin populations venturing out into less productive savannah habitats would have been even more stretched.

Third, using the models to project what will happen as climate warming kicks in over the next century indicates that the apes and the Old World folivorous monkeys (the African colobines and the Asian leaf monkeys) will all go extinct because they lack the dietary plasticity to cope with rising temperatures. In effect, the African apes are almost at the limit of their ability to adjust behaviourally or dietetically to climate change. The orangs represent the limiting case in this respect: they have reduced foraging group size to its absolute limit (solitary foraging) as they have been pushed progressively equator-wards over the last 10,000 years or so by post-Ice Age climate warming. As a result, they are so close to the limits of survival that they are in imminent danger of extinction (even without the helping hand of human-generated deforestation). Hominins would have faced the same problems as the climates dried through the Pliocene and they were forced to venture into more open habitats.

Fourth, a mapping of chimpanzee distributions onto those for their major predators (the lion and leopard) demonstrates that their current biogeographic distribution is limited by predation: they can cope with one or other of the African big cats, but not both at the same time. The presence of both cats prevents chimpanzees from occupying large areas in Angola and the southern Congo that, in terms of time budgets, they should have no trouble living in. This should remind us that predation risk is important even for larger-bodied primate species, and is thus likely to be relevant to early hominins.

I raise these points here to emphasize how crucial time is for animals. Pushed by local climate to spend more time on one activity, animals can easily find themselves having to devote more time than they actually have in the waking day to the essential activities of survival. If they cannot effect savings somewhere, they simply won't be able to survive in that habitat. These are not issues that can be lightly dismissed: they are central to the whole question of whether a species survives or goes extinct.

The last two chapters provide us with both the framework to which we can pin each of the fossil hominin species as well as modern humans *and* the tools we need to assess how much pressure these species were under in terms of their time budgets. By first understanding the magnitude of the problem that each species faced we will be in a better position to ask what options they had for resolving their time budgeting problem and how likely it was that they opted for any one solution. Each of the following chapters will take one of the phase transitions that we identified in Chapter 1, ask what the social brain relationship tells us about the typical community size of the species concerned, and then use the time budget models to ask what they needed to change to have been able to survive where they did.

CHAPTER 4

The First Transition

THE AUSTRALOPITHECINES

By far the most iconic moment in the story of human evolution is represented by a 35 m trackway of fossilized footprints that were discovered at Laetoli in northern Tanzania by the fossil hunter Mary Leakey in 1978. Dated to around 3.6 million years ago, they show the tracks of two adults and a juvenile preserved in light volcanic ash that was falling from nearby Mount Sadiman during one of its periodic eruptions. A light rain hardened the ash shortly afterwards and preserved the footprints under further layers of volcanic debris The two adults were walking more or less in each other's footsteps, while the juvenile's prints cross in and out of their tracks. The footprints are all but identical to those made by living humans, and quite different from those made by an ape. The stride is relatively short and the prints full-footed, indicating that the individuals were walking at a leisurely, unhurried pace. At one point, their tracks are crossed by the prints of an early African horse. It is impossible not to see this as a vignette of everyday family life frozen in time, and many have been tempted to do so.

In fact, by the time these footprints came to light, we had had half a century's worth of australopithecine fossils to play with. Some 54 years earlier in 1924, the South African anatomist Raymond Dart was searching through a box of fossils that had been found by quarrymen near the small town of Taung when he came across the broken skull of what he at first thought was a fossil baboon. Closer examination revealed something completely new: it was neither monkey nor ape, but a seemingly primitive form of hominin. The fossil came to be known as the Taung child, the first of the australopithecine 'ape-men' to be discovered. The number of fossils discovered in sub-Saharan Africa increased exponentially over the ensuing decades, giving us an ever more complex picture of our ancestral tree over this first crucial period of our evolutionary history. However, the main consequence of these discoveries in southern Africa was to reorient our entire picture of human ancestry from the long-held assumption that we had evolved in Asia and place Africa firmly centre stage as the cradle of human evolution. Since then, nothing has emerged from the archaeological record to change our minds on this.

The transition from the Last Common Ancestor to the australopithecines in fact probably took several million years to effect. Nonetheless, by 4 million years ago at the latest, we have a very distinctive lineage of bipedal apes in Africa, with well-established concentrations in eastern and southern Africa (Fig. 4.1). This lineage proved to be so successful that, over the ensuing 2 million years, there were often several species in existence at the same time (Fig. 1.2). The australopithecines were mainly adapted to relatively well-watered woodland and woodland edge habitats beyond the great tropical forests that continued to remain home to the other

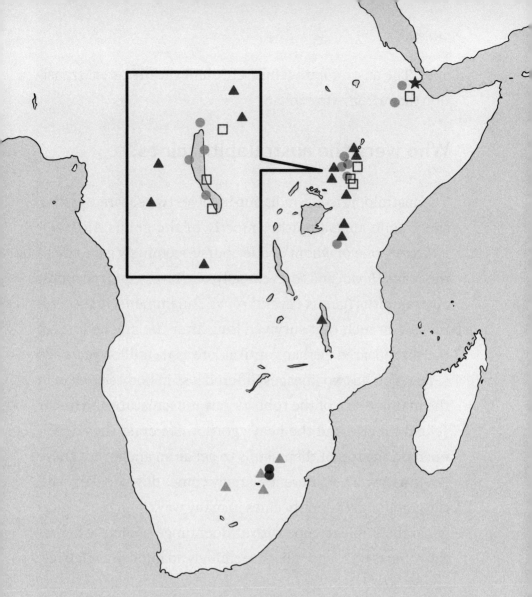

Figure 4.1

Approximate distribution of the main australopithecine sites.

▲ *Paranthropus boisei* and *P. aethiopicus;*

▲ *P. robustus;*

□ *Australopithecus anamensis;*

● *A. afarensis;*

● *A. africanus;*

★ *A. garhi.*

After Klein (2000).

apes. The australopiths thus represent the first great transition that we have to explain.

Who were the australopithecines?

The australopithecine radiation involves two separate grades, the gracile australopiths (mostly of the genus *Australopithecus*), one of whom in due course seemingly gave rise to the genus *Homo* (and so, eventually, ourselves), and the robust australopiths (genus *Paranthropus*), an immensely successful side branch that survived long after the gracile lineage had disappeared, perhaps until as late as 1.4 million years ago (Fig. 1.2). The two lineages differed less in body size than in the massive size of the robusts' jaw musculature and molar (cheek) teeth, and the heavy gorilla-like crest they developed on the top of their skulls to act as an anchor for these jaw muscles. The difference largely comes down to diet, with the gracile australopithecines in many ways being ecological generalists able to cope with a wider range of diets, whereas the robusts were specialists, seemingly adapted to a relatively coarse diet.

The most striking aspect of the australopithecines as a group is not so much how they differ from the other great apes, but how they do *not* differ from them. First and most importantly, their brains were not much larger than those of the chimpanzees (see Fig. 1.3) and, aside from the shift in habitat and their large molar teeth, the only substantive evolutionary development that distinguishes them is bipedalism – notwithstanding the fact that all the great apes *can* walk on two legs (some better than others) and from time to time do so.

Compared with the living great apes, the australopithecines exhibit a very distinct shift from an ape-like skeletal design of long arms and short legs to a more human-like combination of shorter arms and longer legs. The ape body shape is really designed to allow them to shin up the massive trunks of forest trees, using the legs as a platform to support the body while they reach up with their long arms to find a secure handhold they can use to haul themselves up further. Australopithecines are more human-like in form with a design for bipedal terrestrial travel, though they did not yet have the very long-legged body shape that would later characterize the genus *Homo*. Nonetheless, they were well enough on the way to full bipedal locomotion to have many of the features now associated with bipedalism, such as a more human-like bowl-shaped pelvis designed to act as a support for the intestines while striding – very different from the more elongate monkey-like pelvis of the great apes – and thigh bones that were angled in at the hips to create the knock-kneed posture necessary for balanced bipedal walking. Their chest is also somewhat more like the barrel shape typical of later humans, though not yet completely human-like, still having the rather flared lower chest typical of great apes. And perhaps most important of all, the foramen magnum is underneath the skull so that the head is held balanced above the erect spine. In all quadrupedal monkeys and apes, the foramen magnum is at the back of the skull so that the head can look forwards while walking on all fours; when we walk on all fours, we find it difficult to do anything but stare at the ground. By far the most spectacular evidence for habitual bipedalism, of course, comes from that trackway of fossilized footprints at Laetoli. Although it is clear that all the australopithecines were still able to climb quite competently in trees (certainly more competently than modern humans), their foot had already begun to be reconstructed into a human-like foot with the big toe in line with the other toes, rather than out to the side to provide grip on trees as it is in monkeys and apes.

Because the robust australopiths are a side branch that went its own way rather than being on the direct line to modern humans, my focus here will be on the early gracile lineages. These were exceptionally successful while their forest-bound ape cousins underwent terminal decline in the face of deteriorating climatic conditions and the contraction of the great African tropical forests. How did the australopithecines manage this? Caroline Bettridge explored this using our time budget models, and much of what follows is based on her work.

The australopithecines' world

To be able to use the time budget models, we need to know which locations the australopithecines survived in, as well as where they did not. We need to be able to determine the climate at each of these sites, plug the relevant values into the time budget model and ask what group sizes the model would predict for each of these sites. If the model correctly predicts where australopiths actually lived and where they did not, then we can conclude that the fossil species had the same dietary physiology and locomotor constraints as the living species on which the model is based. It is, of course, both important and comforting that our models correctly predict what we actually see in the world, but in fact, in science, the most interesting cases are always those where a model's predictions turn out to be wrong – or, at least, not quite completely right. These cases allow us to ask which equations we need to adjust – and by how much – in order to get the model to predict what we know was actually the case. This second step is the crucial one: it enables us to explore the physiological and anatomical space that australopithecines actually occupied so as to identify just what kinds of novel adaptations they had evolved in order to cope with the new habitats they were trying to invade.

In this respect, it is important to appreciate that the aim of using a particular time budget model – say, a chimpanzee model – is not to claim that fossil hominins (or even fossil great apes, for that matter) *were* chimpanzees, but rather to use the model as a benchmark against which to test various alternative hypotheses about how a fossil species might have behaved, given a set of assumptions about its physiology. In effect, we are using our theories as precision tools to pare away the inevitably messy superficial layers of the real world and get to the heart of what is actually going on below the surface. The more precise the predictions that a model makes, the better this process will work. In evolutionary biology, this approach is known as reverse engineering, and it has proved to be a valuable way of unpacking the evolution of adaptations. Philosophers of science sometimes refer to it as 'strong inference' because it makes it possible to be much more certain about the conclusions we draw, mainly because by requiring our theory to provide a quantitative – rather than merely qualitative – fit to observation our procedure is much more demanding than more speculative approaches.

One final thing we need to take into account at this point is the social brain relationship. Fig. 3.3 tells us that australopithecine community sizes were no bigger than those of living chimpanzees, so they wouldn't have needed any extra social time compared to chimpanzees. The only differences in time budgets from the chimpanzees are therefore likely to be just those in respect of foraging or resting time. That helpfully simplifies our task for this first phase.

In terms of time budget models, the first question we might ask is whether australopithecines are ecological apes or ecological baboons, both models having been proposed for early hominins. Great apes and baboons are constrained by very different aspects of their time budgets (the great apes are limited primarily by the costs of travel, whereas baboons are limited mainly by feeding time) and this provides us with a neat comparison. In part, this reflects a physiological difference between the two taxa:[1] baboons, like all Old World monkeys, can process unripe fruit, but apes cannot. This obliges apes to travel further in search of food patches offering ripe fruit, and so causes their travel time requirements to explode as foraging group size increases and habitat quality deteriorates, as we saw in Fig. 3.7. In contrast, baboons need to move less often because their capacity to exploit unripe fruit allows them to spend more time in resource patches. They are also much less impeded by their skeletal design than apes are: the baboons' relatively longer legs and feet are better adapted for fast terrestrial travel. It is these differences that largely account for the very different distributions of these two genera, and especially for the African apes being confined to a very narrow latitudinal band either side of the equator where tropical forests and their soft fruit occur.

While it is possible to obtain a picture of the climate at fossil sites from the vegetation record (notably from fossilized pollen grains), this does not provide the exact values for the climate indices that we need for the time budget models. The conventional way to determine climate in this detail has been to match the fossil mammal profile of the fossil site to that of a contemporary site from the same continent, and then use the climate data from the contemporary site. The logic is that individual species of animals have very distinct niches that are tied to particular vegetation types, and hence to particular climate regimes, and that these are unlikely to have changed over geological time (i.e. a given species will always have had basically the same ecological niche). The bovids (antelope family) usually provide the best data for this because different species have very different habitat requirements and many of the living families can be found at fossil sites.

Bettridge used the climate data for all the sites at which australopithecine fossils have been found and a selection of control sites where they did not occur (but where other primates, mainly baboons, did so we know the habitats weren't completely primate-unfriendly). Using these values in the chimpanzee model demonstrates rather starkly that, had the australopithecines been conventional chimps, their predicted community sizes would have been embarrassingly small (typically well below 10 animals) at all the South African sites (Fig. 4.2). In other words, they could not have survived in habitats where we know they were actually very successful. They would even have been hard pressed to maintain communities of any significant size at any of the sites after 2.5 million years ago when the climate cooled significantly and things should have been less stressful for them. Even then community sizes would typically have been less than about 15 individuals, well below the viable community size in modern chimpanzees (for whom a minimum community size of 40 is required). The time budgets for these sites show that it was travel time costs in the relatively dry habitats that australopithecines typically occupied that were the problem. So the bottom line is that, whatever else they might have been, they weren't ecological chimpanzees. In contrast, the baboon model, scaled up for the australopiths' larger body size, has australopithecines living all over sub-Saharan Africa in locations where there is no evidence that they had ever occurred. So they weren't ecological baboons, either.

Although the ape model predicts community sizes of zero for most australopithecine populations, in fact the time budgets predicted by the model for those habitats where the australopithecines actually occurred are not too far adrift from the equilibrium point. The average values predicted for the four key activity categories are shown in Fig. 4.3. These add up to just 107 per cent and it is precisely this 7 per cent overspend that prevents them devoting enough time to social grooming to maintain communities of the size their brains tell us they had. While finding savings of 7 percentage points from the other activity categories is always going to

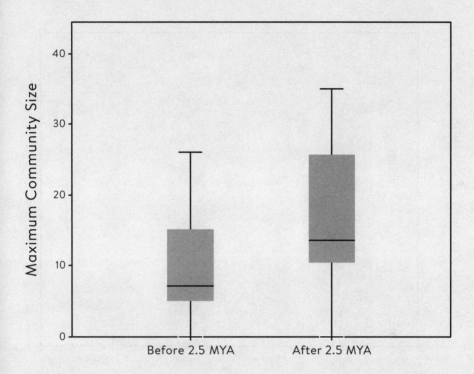

Figure 4.2

Median (with 50 and 95 per cent range) maximum community sizes for early (before 2.5 million years ago (MYA)) and late (after 2.5 MYA) australopithecine populations, as predicted by the chimpanzee time budget model. Source: Bettridge (2010).

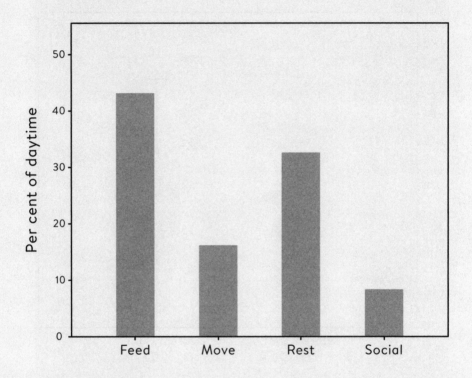

Figure 4.3

Time budget allocations for australopithecines, based on the Bettridge (2010) model for sites at which australopiths are correctly predicted to live.

be challenging, this level of overspend is modest enough to be achievable without having to compromise on community size. They obviously did manage this because we know they lived in these habitats. So how did the australopithecines solve their time budgeting crisis?

Was bipedalism the solution?

One possibility is that the solution lay in some advantage to bipedalism, since this strikingly novel adaptation was obviously acquired very early on and in many ways defines the australopithecines. By lengthening their legs compared with apes, early hominins might have made it possible to travel further per stride, and so use less time (and energy) than great apes do to travel a given distance (or, alternatively of course, travel further in the same amount of time).

Physiological studies of locomotion show that, in terms of oxygen consumption, bipedalism is slightly *less* efficient than quadrupedalism for chimps, mainly because they have to walk with bent knees. This is because their elongated ischia (the wings of the pelvis to which the leg muscles attach) get in the way when they walk bipedally. The remodelling of the hominin pelvis into the bowl shape that we now have allows our legs to swing without obstruction, enabling us to stride with straight legs. In addition, modern humans gain a 75 per cent advantage in energy savings over chimpanzees when walking bipedally due to a number of unique anatomical adaptations to the hip and foot that provide elastic propulsion. The cartilages that link the bones of the arched human foot, for example, act like a spring and store energy that gives extra propulsion on each stride. Since australopithecines weren't as fully bipedal as modern humans, the default would be to assume that the only benefit they gained was from increased stride length – or, at least, that this is the benefit needed to get the trait off the ground, allowing the adaptations to the hip and foot to kick in later. Robert Foley and Sarah Elton modelled the energetic costs of bipedal vs quadrupedal travel in trees and on the ground and concluded that bipeds gain an advantage over quadrupeds when more than about 65 per cent of the time is spent on the ground. The switch to a more terrestrial lifestyle thus seems to have been critical.

Australopithecine females are estimated to have had a leg length of about 52 cm, compared to 44 cm for female chimpanzees. While not in the same league as the gain for *Homo ergaster* females (whose leg length was around 80 cm, an increase over chimps of 81 per cent), it would nonetheless have been enough to reduce the 16.4 per cent travel time predicted by the chimpanzee model for sites where australopithecines did occur to 14 per cent, a saving of roughly 2.5 percentage points or about a third of what is required. Less time spent travelling would also mean some saving on feeding time to fuel travel, perhaps increasing the total saving to 3 percentage points. (I will assume that this saving was not absorbed by the need to travel further in the more open habitats favoured by australopiths. There is no evidence to suggest that the australopithecines were as nomadic as later *Homo*.) Taking this gain at face value, that might well have been enough to get bipedalism up and running, but it would still have left another 4 percentage points to find to get australopithecines living in their most benign habitats – and a bit more than that to get them living in the more marginal ones.

One additional advantage to bipedalism is cooling. In relatively open habitats, the body of a quadrupedal animal absorbs more sunlight than a bipedal one because standing upright means that only the top of the head and the shoulders are exposed to the sun, especially during the middle period of the day when the sun is overhead and at its hottest. A quadruped will thus overheat more quickly than a biped. The brain has very narrow tolerances temperature-wise: raising the temperature of the brain by more than 1°C results in heatstroke and, within a relatively short period of time, brain cells start to die. By minimizing the amount of radiant heat absorbed by the body from the sun during the middle of the day, a bipedal animal might be able to remain active longer when the sun is at its hottest. The major issue here is resting time, since all mammals have to take shelter when temperatures get too high, and a reduction of 4 percentage points in resting time through animals being able to remain active slightly longer into the middle of the day would be more than enough to solve the remaining time budget overshoot in Fig. 4.3.

Modern humans have two features that are not only unique among the primates but also seem to be directly related to this heat-load problem, namely the loss of fur over most of the body (other than the head and, to a much lesser extent, shoulders – the areas of the body most exposed to the sun at midday) and a greatly increased capacity for sweating (we have many times the number of eccrine sweat glands in the skin than all other primates except for baboons, the only other terrestrial open country species). Physiological models developed by Peter Wheeler suggest that reduced exposure to direct sunlight combined with evaporative cooling through sweating would have enabled a naked bipedal hominin both to remain active longer than a quadrupedal one and/or to travel twice as far on a litre of water. The key point here is that sweat evaporating off fur just cools the tips of the hair and not the skin underneath; to benefit from evaporative cooling of sweat, the animal has to be naked.

Wheeler's thermal load model has recently been challenged. The biologists Graeme Ruxton and David Wilkinson pointed out that walking itself generates heat, and this internally created heat needs to be added to that produced by sunlight striking the body. With this additional internal source of heat, they argued, bipedalism gives only a small advantage, and the main gain would come from hair loss and sweating. A hairy australopithecine (whether bipedal or quadrupedal) would have been unable to survive in open habitats because the combined thermal load would have been more than it could dissipate; even a hairless animal would be unable to lose heat fast enough to offset overheating when active during the middle of the day. This is illustrated in Fig. 4.4. The filled circles show the heat load, and the horizontal line indicates the amount of heat that could be dissipated by evaporative cooling from the skin (approximately 100 W per hour). Under the Ruxton–Wilkinson model, heat load would exceed the heat that could be dissipated from roughly 7.30 a.m. until 6 p.m. This implies that, on their own, the thermal benefits would not have been sufficient to favour the

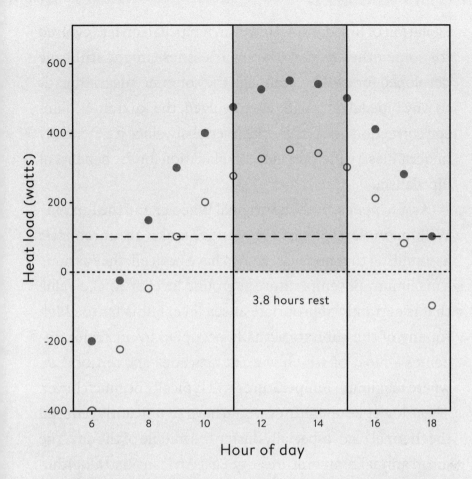

Figure 4.4

The Ruxton–Wilkinson model of heat load assumes that early hominins were travelling all day. The solid symbols are their estimate of heat load for a hairless biped, and suggest that for most of the day australopiths would be generating and absorbing far more heat than they could dissipate (indicated by the horizontal dashed line at 100 W). The open symbols show the effect that the lower temperatures faced by australopithecines at the sites where they actually lived would have had on heat load: they would have been able to be active significantly longer in the morning and evening. The Bettridge model estimates that australopithecines would have spent 3.8 hours resting each day; like all open-country monkeys and apes, they would have done this during the middle of the day when temperatures were highest (the shaded area).

evolution of bipedalism. However, if bipedalism had evolved for some other reason, then hairlessness might still have developed for cooling. Although they offer no suggestions as to why bipedalism might have evolved, the Ruxton–Wilkinson correction needs to be taken seriously, since it potentially undermines a widely accepted explanation for the benefits of bipedalism.

As it happens, both the original Wheeler and the Ruxton–Wilkinson versions of this model make an unrealistic assumption that no one seems to have noticed: they assume a maximum air temperature at ground level of 40°C, a value that is certainly appropriate at sea level but is far too high for any of the habitats actually occupied by australopithecines – most of which were at altitudes above 1,000 m, where maximum temperatures will typically be much lower. These lower temperatures would have significantly reduced the thermal load, especially during the middle of the day. The mean annual temperature at 35 East African australopithecine sites was 25°C and at five South African sites it was a mere 20.4°C,[2] compared to an average value over the day of 32.5°C assumed in both the Wheeler and Ruxton–Wilkinson models. These lower temperatures would reduce heat load in the middle of the day by around 200 W, which would mean that the australopiths could have been active for another 2.5 hours each day (as indicated by the open circles in Fig. 4.4). In other words, they could have been active for as much as four hours each day during the early morning and evening without exceeding their thermal load threshold. I will return to this point later.

First, let us consider an alternative suggestion that has

attracted some attention in recent years, namely the suggestion that bipedalism allowed early hominins to carry food so that they could eat it later in a location with less risk from predators. Recent observations of both chimpanzees and gorillas wading through water-logged swamps in the northern Congo have even rekindled interest in the idea that bipedalism may owe its origin to wading in lake or seashore shallows in search of food. However, while occasional bipedalism clearly has an advantage in this respect, it is not entirely clear why *habitual* bipedalism would then be so advantageous. Chimpanzees do quite well enough carrying food bipedally on the odd occasions when they really need to (usually when raiding crops), but it is difficult to envisage any circumstances under which it would actually pay the animal to carry food *all the time* so as to make habitual (as opposed to casual) bipedalism so necessary. If anything, carrying is more likely to have been a consequence of bipedalism, not its evolutionary cause. More importantly for us, it is not clear how carrying would help solve the early hominin time budget crisis, since it would have no effect on either feeding or travel time as such.

If the major advantage from bipedalism comes from the cooling effects of sweat evaporating off bare skin, the issue is whether the australopithecines were furred or not. Wheeler assumed that they were not, but others have argued that hair loss only occurred with the appearance of *Homo* around 1.8 million years ago, even though we know that bipedalism evolved a great deal earlier (i.e. by 5 million years ago at the latest). Irrespective of whether the second view is correct, we still have a residual time budget crisis for australopithecines to sort out since hair loss seems to influence mainly the distance and timing of travel rather than the time allocated to travel per se. The crucial question is whether the australopiths could have afforded to devote twice as much time to travelling if sweating allowed them to travel twice as far. The time budgets suggest not. A serious possibility is that hair loss did come in later with the appearance of the more nomadic *Homo ergaster*.

Dietary solutions to the time budget crisis

The solution that living chimpanzees use to solve their time budgeting problem is to reduce their foraging group (as opposed to community) size. This reduces their travel time costs significantly (Fig. 3.7), and allows them to devote more time to feeding and social interaction. However, chimpanzee foraging groups are already down at around 3–5 – and at the lower end of this distribution in precisely the kinds of open woodland habitats occupied by the australopithecines. However, there is little more time that can be saved here. Even if foraging party size is reduced to its limit at one (i.e. solitary foraging), the gain in terms of saved travel time is minimal: Fig. 3.7 suggests that a reduction in foraging party size from 5 to one would yield a saving of just 1.4 percentage points on travel time – far too little to be of much help. In any case, there are predators to worry about on the African savannahs. Orangs can forage solitarily because they have few predators in their current habitat and plenty of trees to escape into. Chimpanzees seem to be unwilling to go to these lengths because serious predators still stalk their habitats.

Australopithecines would likely have faced more serious predators in larger numbers – and had fewer trees available as refuges – so the suggestion that they might have opted for the orang strategy is implausible, even if the savings in travel time made it worthwhile.

One possibility might be that australopithecines found ways to cut down on time spent feeding by changing to a more efficient diet – one that needed less time for processing, was richer in terms of nutrients, or perhaps occurred in denser patches that necessitated less travel. We can tell quite a lot about the diets of fossil species, in part from the size and shape of their teeth, and in part from the scratches and abrasions that chewing food leaves on the teeth. The australopithecines have large chewing teeth (and they get larger still in the robust australopiths), suggesting a relatively coarse or hard diet that needed a great deal of chewing. Under high resolution electron microscopes, the marks on their teeth show the kinds of damage suggestive of a heavily abrasive diet. This might be a consequence of a hard, brittle diet (hard nuts, for example) or a great deal of grit in a diet collected from the ground, or both. Grit would be suggestive of feeding on underground plant parts (roots, tubers, rhizomes) that have to be dug up.

In recent years, chemistry has helped us out here, too. The carbon atom has two isotopes, C^{13} and C^{14}, that give rise to two quite distinct metabolic pathways, known as the C3 and C4 pathways. It turns out that the C4 pathway is found only in tropical grasses, sedges, succulents[3] and some of the beet family, while the C3 pathway is characteristic of shrubs and trees (and hence characteristic of ape and monkey diets based on fruit and nuts). Since the carbon atoms in food are used to build the body, measuring the ratio of carbon isotopes in the bones of fossil species tells us about their diets.

Analysis of carbon isotopes from the teeth of the later australopithecines suggests that their diets were rather more C4-based than we would expect for an Old World monkey or ape. Ethiopian *Australopithecus afarensis* from around 3.2 million years ago have a C4 signature that is midway between that of grazers (equids,[4] hippos, warthogs) and that of browsers (giraffes) from the same habitat, indicating partial rather than full commitment to a C4 diet. Moreover, their C4 signature seems to be quite variable across time, suggesting that their use of these resources were either seasonal or sporadic. In contrast, chimpanzees and the early australopithecines (*Ardipithecus ramidus* and *Australopithecus anamensis*) exhibit hardly any C4 signal, despite apparently living in C4-plant dominated habitats (woodlands). So *afarensis*'s exploitation of C4 plants seems to represent a new dietary trend. This trend appears to have been taken to its extremes by the later robust australopiths, in particular East African *Paranthropus boisei*, whose diet was very C4-biased. It seems that at least the later australopithecines were doing something very different from the classic African apes.

There are two possible ways in which an animal might acquire a strong C4 signature: one is by eating C4 plants, and the other is by eating animals that eat C4 plants. There is evidence of cut marks on bones, at least from the later australopithecine period, and these have been interpreted as indicating that australopithecines acquired and ate meat – perhaps by scavenging from the kills of large predators (lions and leopards, and maybe hyenas). However, raw mammalian flesh is not so easily digested by apes (including humans), and, at least in humans, protein poisoning can result when the diet is too meat-rich. Cooked meat is another matter, of course, as cooking increases the digestibility of meat by about 50 per cent. However, since control of fire is *most* unlikely at this early date (see Chapter 5), it is unlikely that any of the australopithecines had diets with a meat content that was significantly greater than that we find in chimpanzees.

Bone-breaking, however, is a possibility, since this gives access to the marrow in long bones – and better still, brains in skulls. Both are much easier to digest than red meat. Bones cease to be of interest to most large predators once they have removed the accessible meat, making them a less hazardous target for scavengers. From around 2.5 million years ago, we begin to find evidence of crude stone tools, known as the Olduwan toolkit after the East African site of Olduvai Gorge where they were first discovered. These are essentially hammer stones with a modest cutting edge, though what they were used for is unclear. However, their shape and weight is perfect for breaking bones to extract the marrow. In fact, there is direct evidence of percussion marks on ungulate bones from Dikika in Ethiopia dated to around 3.4 million years ago that have been interpreted as attempts to get at marrow, so maybe this is why they developed this toolkit. Chimpanzees make ready use of tools of exactly this kind to break open the hard casings of the nuts of several tree species, including the Guinea palm and *Coula edulis*. Although these nut-bearing trees do not grow in the more open savannahs and woodlands of East Africa, it would have been but a very small step from using these tools to crack nuts to using them to crack open long bones or skulls. This might have made all the difference in terms of survival in the more marginal kinds of habitats that australopithecines were beginning to occupy, especially as the climate became drier in the lead-up to 2 million years ago when the climate in Africa rapidly started to cool and dry. Would marrow and brains have been enough to reduce australopithecine feeding time to the extent needed? Perhaps, but the real problem is that if tools provide the evidence for marrow extraction, then even at 3.4 million years ago it would seem to have been too late in time to have been more than a late transitional phase. It would not have solved the time-budgeting crisis of the early australopithecines.

An alternative suggestion is that termites may have been responsible for the C4 signature. Termites are grazers and live in very large nests, which make them a rich food source. These nests are made of mud that is so concrete-like as to be virtually impenetrable without the equivalent of a jack hammer. Chimpanzees certainly eat termites when they can – and relish them. But they do so by the rather laborious means of fishing for them: they poke grass stems into the entrance holes of the nest and eat the termites that cling to the stem in a misguided attempt to repel the intruder. It's slow work at best, and unlikely to offer any serious savings in time. Like chimpanzees, australopithecines surely exploited termites during their annual flights to establish new nests, but I remain unimpressed by the extent to which their limited technology would be any better than the chimpanzees' at giving them access to large numbers of insects in the nest sufficiently often to make a significant contribution to their diet.

A fourth suggestion is that the australopithecines began to exploit the underground storage organs (roots and tubers) of certain families of herbaceous plants, many of which have the C4 metabolic pathway. There is, of course, no direct evidence for this: vegetable foods don't fossilize. Nonetheless, the fact that many plants that have these kinds of underground components are C4 and occur in lake margins where australopithecine fossils are often found at least provides a smoking gun for this idea. And, as we have already noted, this is given added weight by the dental evidence for a relatively abrasive diet indicating the exploitation of underground foods.

In sum, there are a number of options on the table that might work. The key question is whether there is any evidence from the time budgets to show that any of these may have been their solution of choice.

What time budgets tell us

In trying to understand what enabled the australopithecines to survive where they did, Caroline Bettridge first ran a sensitivity analysis of the feeding, resting and travel time equations in the chimpanzee model. She did this by altering the steepness of each equation in steps to see what effect it has on the model's ability to correctly predict where australopithecines did and did not live. Changing the feeding and resting equations made hardly any difference, but changing the steepness of the moving time equation had a dramatic effect, reducing the number of incorrect predictions for the australopithecines from 30 per cent to as little as 8 per cent of sites. This reinforces the suggestion that, as with all great apes, it was travel time (and hence the distances that had to be travelled between food patches) that was the problem for the australopithecines.

As a check on this, Bettridge substituted the baboon moving time equation into the chimpanzee model to see what effect a baboon-like resource distribution would have had. Using the baboon moving time equation, adjusted for the larger body mass of the australopithecines, with a minimum community size of 10 (the absolute lowest ever observed in chimpanzees) improved the chimpanzee model's ability to predict the presence or absence of australopithecines from 26 per cent (for the basic chimpanzee model) to a very respectable 76 per cent. Most importantly, it correctly predicts that australopithecines could live in all the places where we know they lived – which the chimpanzee model conspicuously fails to do. It seems that whatever australopiths were doing, they were not foraging like apes but more like baboons.

What really seems to make the difference in the model is the effect of rainfall, and this perhaps provides us with a clue as to what australopithecines might have been up to. Remember that their diet had a high C_4 signature, with one suggestion being that this was due to a diet based on roots and tubers. Their fossils seem to have been particularly associated with lakeside and riverine habitats – exactly the kinds of places that make it possible for baboons to survive in more open savannah habitats that would otherwise be too dry even for them to cope with. Now, as far as the time budget models are concerned, the effect of being near large bodies of water is exactly the same as increasing rainfall: both result in more water being available in the ground for plant growth. The result is permanent strip forests along watercourses in savannah habitats that would otherwise be too dry to support forest. Such habitats are also commonly associated with flood plains: flat areas either side of a large river or around a lake that flood during the rainy season, creating a rich microhabitat that is perfect for succulents and other plants that store nutrients in roots while waiting for water to become available. A switch to underground storage organs would certainly have allowed australopithecines to live in these kinds of habitats without having to alter their digestive physiology too much, just as baboons in fact do in exactly these kinds of habitats. At the same time, it would have dramatically reduced the need to travel far since these patches tend to be rich and so able to support large numbers of individuals. That might even have allowed australopithecine communities to forage as a single group rather than having to adopt fission–fusion sociality, and so allowed them at the same time to cope with the higher predation risk that they would have faced away from the trees.

A somewhat similar response has been noted among gorillas in the *bai* habitats of the western Congo. *Bais* are open patches in the swampy forests of this region, often associated with pools of water, which attract several gorilla groups to share their rich food resources. Gorillas occupy a very different kind of habitat to the australopiths, but the ecological effect is similar: groups that would normally have to forage separately can forage together when the resource patch is rich enough and large enough to allow many animals to feed there for a long period of time without getting in each other's way. A similar response can be seen in gelada baboons, who as grazers are able to congregate in very large numbers (up to 500 individuals) on rich grassy swards – and can do so even in the dry season (when grass leaves are dry and inedible) by digging for roots and rhizomes. The gelada are a special case among the primates and they live in a very special kind of habitat (the high altitude grasslands that only occur between 1,700 and 4,000 m above sea level in eastern Africa), but they illustrate the same principle.

There is one last point to consider. Although the patches of gallery forest and woodland associated with lakes and large rivers offer shade, the areas of woodland and open flood plains beyond would not. Australopiths would have been forced to rest during the middle of the day when temperatures were at their peak, as do most open country species like baboons, who typically rest in whatever shade they can find for several hours around midday. The chimpanzee model predicts that australopiths would, on average, have had to devote almost exactly a third of their day to resting (3.8 hours) and only a very modest proportion (about 1.9 hours) to travelling in the habitats they actually occupied. If Ruxton and Wilkinson are right in concluding that early hominins would have been restricted to travelling only very early and late in the day, then the australopithecines could easily have fitted two hours' travelling into the morning and evening periods of the day according to their model, and still have had plenty of time free around the middle of the day for a 4-hour siesta (Fig. 4.4). In other words, early hominins would have had to do most of their travelling to find a new feeding site in the early morning. Once they had arrived at a site, they could spend the rest of the day there, moving again late afternoon to a safe night-time refuge in the gallery forest nearby. Baboons have exactly this pattern of activity in just these kinds of habitats. It rather looks like a radical change of diet wouldn't have provided the solution to their problems, even though meat may have made a small contribution as it does with chimpanzees. But a unique habitat associated with proximity to large bodies of standing water might well have made a difference, especially if they also switched to novel kinds of underground plant foods. Bipedalism seems to have provided a crucial cushion in terms reducing some thermoregulatory costs, but more importantly perhaps by allowing them to move more quickly and efficiently. One implication of this is likely to have been that australopithecine communities would have been quite patchily distributed in small clusters, often separated by considerable distances, along narrow riverine forest corridors.

Most of the South African australopithecine sites are associated with limestone caves edging river valleys. Although a convincing case has been made for at least some of the fossils being the residue from the kills of leopards or birds of prey – falling from roosts in trees down sink shafts below – it is nonetheless likely that caves provided important night-time refuges for these species. Caves have two properties that would have made them important as night-time refuges in these habitats. One is as a place of safety from predators in areas where large trees in which to roost are in limited supply. The other, and perhaps more important, is that caves dramatically increase the temperature, especially at night when temperatures can be very low at these southern latitudes. Even at the equator, night-time temperatures can be as low as 10°C at altitudes above 1,000 m (as I know from experience during field work). In a study of feral goats in Scotland, we found that minimum temperatures were 3–5°C higher in caves that the animals used as night-time shelters than outside the caves. Similarly, a study of cave use by baboons in southern Africa found that night-time temperatures inside the caves were about 4°C warmer than those outside.

When the air temperature is below an animal's body temperature, it has to use energy to keep warm, and this costs extra feeding time. The lower the temperature, the worse the problem gets. As with the goats I studied in the northwest of Scotland, caves may have been crucial to the survival of early hominins in southern Africa. The ape (but not the baboon) time budget models are based on equatorial sites and do not include any adjustment for the cooler climates of southern Africa as such. Although we cannot say whether the use of caves simply offset higher thermoregulatory costs in southern habitats or whether it allowed additional savings on their feeding time budget, caves almost certainly played an important part in allowing australopithecines to colonize southern Africa as successfully as they did.

Australopithecine social life

So far, we have established that australopithecines must have lived in communities of about the same size as those of modern chimpanzees (the social brain tells us so), but may not have needed a fission–fusion social system that was quite as extreme as that of the chimpanzees. Despite the fact that their size would have buffered them somewhat against terrestrial predators, large social groups would still have been their main defence against predators on open pans and flood plains. What kind of social system did they have?

There has been a view, first articulated as long ago as 1981 by the American palaeoanthropologist Owen Lovejoy, that monogamy evolved early in the hominins, largely on the grounds that two-parent care was essential for raising our altricial, large-brained offspring. Lovejoy originally argued that monogamy must have been characteristic of all members of the genus *Homo*, and so universal among the later hominins from around 2 million years ago. This seems like a plausible claim since this timing would coincide with the first significant uplift in hominin brain size (Fig. 1.3), a natural point at which two parents might be necessary for rearing offspring. Lovejoy has, however, since extended this claim back to the earliest known australopithecine, *Ardipithecus ramidus*, dated to around 4–5 million years ago. On this view, monogamy, like bipedalism, would be a universal trait of all hominins.

Inevitably, it is tempting to see the Laetoli footprints – two adults and a juvenile moving in close coordination – as supporting evidence for this claim: what could have been more natural than a bonded pair, accompanied by their offspring, travelling, somewhat chimp-like, apart from the rest of the community? But we have to be sceptical of this interpretation. For one thing, the area that has been excavated is very small and we don't really know how many other individuals were nearby – perhaps the three were not actually that far from the rest of the group. Baboon groups often get dispersed over as much as half a kilometre in these kinds of habitats. There are, however, two more serious reasons to be sceptical. First, chimpanzee males and females often form temporary consortships when the female is sexually receptive and separate themselves off for a day or two from the rest of the community to avoid being disturbed, but no one would ever suggest that the uncompromisingly promiscuous chimpanzee is monogamous on the basis of a chance observation of a single consort pair. Second, given that there had been no increase in brain size across this first transition, why would australopithecines have needed biparental care at all? If chimpanzees can cope happily without need of biparental care, why should australopithecines with their similar-sized brains have suddenly struggled?

In Chapter 2, we saw that monogamy evolves only as a response to infanticide, with biparental care emerging afterwards. So if the australopithecines were genuinely monogamous, it must have been because they faced a significantly higher risk of infanticide than the chimpanzees, forcing the females to opt for a bodyguard strategy. The Harcourt–Greenberg infanticide model (summarized in Chapter 2) indicates that infanticide risk isn't high enough in chimpanzees to cause females to seek out hired guns for protection. However, with the parameter values that characterize gorillas, infanticide risk is much higher; with males that are significantly larger than females (and hence greater power differentials between males), females opt for the bodyguard strategy. Since brain size determines both community size (and hence the number of males in the group) and reproductive characteristics (such as the length of the reproductive cycle), there would have been little to choose between the australopithecines and chimpanzees on either of these key traits. More importantly, the analyses of the evolution of monogamy in primates that we discussed in Chapter 2 clearly suggest that monogamy is a one-way street: there is no backtracking to polygamy. Monogamy is also associated with the dispersal of mated pairs into separate territories – and a consequential radical reduction in group size for which there is no evidence from australopithecine brain sizes. More importantly, in primates, monogamous species always have smaller brains than polygamous species.

As it happens, anatomy comes to our rescue here, and offers us three sources of indirect information on australopithecine social life. These are: (1) a limited amount of data on anatomical indices of polygyny, (2) rather more extensive evidence on sex differences in body size, and (3) trace metal signatures in teeth that tell us something about dispersal patterns.

The ratio of the second to the fourth fingers (the so-called 2D:4D ratio, the D standing for 'digit') is influenced by the levels of testosterone to which the foetus is exposed in the womb. In promiscuous species of primates, males have to fight it out with each other to gain access to females and as a result both males and females have higher testosterone levels than is typical of monogamous species. Consequently, in polygamous species, the males are characterized by low 2D:4D ratios (they have short index fingers), whereas in monogamous species this ratio is close to equality. Among the living apes, only the gibbons (the so-called lesser apes) are obligately monogamous: all dozen gibbon species live in separate pairbonded groups of one male and one female, with their dependent young. In contrast, all the great apes have promiscuous (chimps and orangs) or polygamous (gorillas) mating systems.

When Emma Nelson and Susanne Shultz applied this finding to two Miocene apes (*Hispanopithecus* and *Pierolapithecus*), and a handful of hominins for which finger bones exist, they found that all but one of the species were about as polygynous as chimpanzees and orang utans (Fig. 4.5). *Ardipithecus ramidus* was clearly chimp-like in its mating strategy, and the five Neanderthals and one archaic human (*Homo heidelbergensis*) in the sample all had digit ratios at the lower end of the modern human range, roughly in line with the gorillas with their harem-based mating system (where males compete less directly with each other, but mate polygynously). The only exception was *Australopithecus africanus*, which fell in the upper quartile of the modern human distribution, but still far below the values for the obligately monogamous gibbon. So at best, *A. africanus* was about as monogamous as modern humans, and, as we will see in Chapter 9, that isn't very monogamous: polygamy is much the most common form of mating system in modern humans, and monogamy

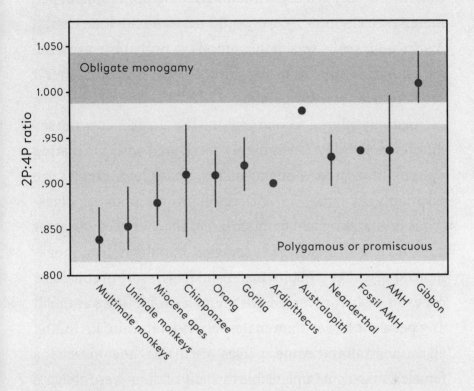

Figure 4.5

Digit ratios (indexed as 2P:4P ratio for fossil species, using the lengths of the finger bones themselves; and as 2D:4D ratios based on finger lengths in living species) for various hominins, great apes (including two Miocene apes) and Old World cercopithecine monkeys (guenons that live in unimale groups, and macaques and baboons that live in multimale/multifemale groups). The dots show the mean value for each species; the whiskers show the range in the data values. The monogamy and polygamy boxes are defined, respectively, by the range for gibbons (the only obligate monogamists) and the range for monkeys and living great apes (all of whom are either polyga-mous or promiscuous). Redrawn from Nelson et al. (2011) and Nelson and Shultz (2010).

occurs only where it is socially enforced. Given the finding that monogamy is a cognitive and demographic sink in primates (see Chapter 2), we would have to conclude that, if A. *africanus* really was monogamous, it probably wasn't the ancestor of modern humans, since all other hominins after it seem to have been polygamous.

Anatomy offers a second angle on this. In primates, the sex difference in body size is highly correlated with the mating system: in monogamous primates, males and females are invariably the same size, whereas in promiscuous or polygamous primates males are much larger than females – in great apes, up to twice as large as females. In obligately monogamous species like gibbons, males and females are of equal size (or even slightly reversed, with females sometimes as much as 5 per cent larger than males). In no hominin, including all the australopithecine species, are males the same size as females, making it implausible that any of them were obligate monogamists. With a male : female body weight ratio of 1.56 (for *afarensis*) and 1.35 (for *africanus*), australopiths were, if anything, more sexually dimorphic than chimpanzees (on average, 1.27) though not as dimorphic as baboons (1.8) or gorillas and orangs (around 2.0). While *africanus* is clearly trending in the same direction as modern humans (our weight ratio is around 1.2), they are still more dimorphic than we are and a genuinely monogamous mating system seems quite implausible.

One final aspect of behaviour that is important is the dispersal patterns of the two sexes. In polygamous mating species, one sex typically remains in its natal territory and the other moves at puberty to a neighbouring community; in monogamous species, both sexes usually move because neither parent is especially tolerant of same-sex offspring once they have matured. Strontium profiles in the teeth of *Australopithecus africanus* and *Paranthropus robustus* specimens at the Sterkfontein site in South Africa provide unexpectedly direct evidence of sex differences in dispersal. Like oxygen and carbon, strontium has two isotopes, and these vary in their frequency in the soils of a particular location. The isotopes are taken up by the plants in proportion to their local availability, and then in turn absorbed into the tooth enamel of the animals that eat the plants. By analysing the strontium ratios in the teeth of a fossil, it is possible to determine whether or not it was living as an adult in the same place as it was born.

For both australopith species present at Sterkfontein, the strontium signatures indicate that the females had moved further than the males, and that they had moved at least 3–5 km (given the distance to the nearest source that had plants with the same strontium signature). Since this pattern of dispersal (males remaining in their natal territories and females dispersing) is similar to that in chimpanzees and perhaps gorillas (and the opposite of that in cercopithecine monkeys), a mating system like that of chimpanzees seems likely for these australopithecines. A monogamous or pairbonded mating system is pretty unlikely.

In sum, it seems implausible that the australopithecines could have been monogamous. If they foraged in small dispersed parties in the way the chimpanzees do, then promiscuity seems the most likely mating system because they would have been unable to defend several females at the same time when the females were so scattered. But if, as I have suggested, they foraged in larger groups on rich lakeside and riverine flood plains, then a form of harem-based polygamy not unlike that found in hamadryas and gelada baboons and, perhaps, gorillas is a possibility. Small harems defended by a male, with the harems of related males foraging together as a community, much as happens among the gorillas of the western Congo *bai* habitats, seems a plausible suggestion. It is likely that males would have had to evolve a greater tolerance for each other to allow several of them to forage together in the large groups that may have been necessary as a predator deterrent on floodplains. If so, this might account for the marked reduction in canine size in australopithecine males.

After a long and successful career, the australopithecines faded out in the aftermath of a major climate change that kicked in around 1.8 million years ago. The climate cooled, and habitats in Africa became drier as more atmospheric water began to be locked up in the polar ice sheets. From about 2 million years ago, the gracile australopithecines that had occupied much of the continent were replaced by the robust forms, *Paranthropus boisei* in eastern Africa and *P. robustus* in southern Africa. These species were characterized by massive cheek teeth, heavy jaw muscles supported by large gorilla-like crests on the top of the skull, and a diet that emphasized C4 species – probably reflecting an increasing consumption of underground storage organs and succulents.

In effect, they made the same adjustment in the face of deteriorating climate conditions as the gorillas were to make much later when these opted for a more leaf- and pith-based diet in poor quality habitats like the Virunga Mountains where they were studied by Dian Fossey. The robusts were successful enough to linger on until as late as 1.4 million years ago, but eventually they too succumbed to the pressures of a deteriorating environment, their disappearance marking the end of a particularly successful phase of hominin evolutionary history. Meanwhile, the second transition had been underway for the better part of half a million years by the time the last of the robusts disappeared. It is to this second phase that our story now turns.

CHAPTER 5

The Second Transition

EARLY *HOMO*

When the famous fossil hunter Louis Leakey found an unusually gracile hominin fossil in East Africa in 1962, he gave it the name *Homo habilis* ('handy man') because it was the maker of the primitive stone tools that seemed to appear around the same time – surely a mark of the fact that they belonged on our side of the 'great divide'. Dated to around 2.3 million years ago, this new species seemed to be a natural bridge between the earlier australopithecines and the Asian and European *Homo erectus* populations that were to follow. However, as more specimens from this and related species like *H. rudolfensis* and *H. ergaster* were discovered in East Africa during the closing decades of the century, it began to seem that *habilis* might just be another australopithecine. So they are now classified as australopithecines.

Nonetheless, the habilines are clearly part of the big story that was just about to break: the appearance of a radically new hominin, represented in Africa by the species *Homo ergaster*. Within a few hundred thousand years of its first appearance, this new species was to dominate the African scene and make the first of many dispersals out of Africa to colonize Eurasia. *Homo ergaster* was characterized by a significant increase in brain size, the remodelling to the skeleton to produce for the first time a recognizably modern human form defined by long legs adapted for a striding gait, a more nomadic lifestyle and what has come to be seen as the iconically definitive stone tool – the Acheulian handaxe. The handaxe and its maker were to persist almost unchanged for well over a million years from their first appearance in East Africa around 1.8 million years ago until more modern tools and their definitively human makers began to appear around 500,000 years ago.

We have, in fact, known about this second phase of hominin evolution for some considerable time. The very first specimens – mainly fragmentary remains of skulls and the odd limb bone – had been discovered in Indonesia during the 1890s, and named *Pithecanthropus erectus* (the erect, or upright, ape-man) by Eugène Dubois, the Dutch anatomist into whose hands the fossils had fallen. Then, in the 1920s and 1930s, a series of very complete skulls and assorted bones were discovered in China, and named *Sinanthropus pekinensis* (the China ape-man from Peking (now Beijing), near where it had been discovered). Unfortunately, the original specimens of 'Peking Man' disappeared during the Second World War while the Americans were trying to rescue them. Luckily, their discoverer, the German anatomist Franz Weidenreich, had published detailed anatomical descriptions of all the specimens, and had made excellent plaster copies of them, a set of which had been sent to the Natural History Museum in New York sometime before the Japanese invaded China in 1937. It is these plaster copies on which our knowledge of this species depended for over half a century. Eventually, all these various Asian specimens (and the additional fossils subsequently discovered in Europe) were absorbed into a single species, *Homo erectus*. Later, *H. erectus* was combined with the African *Homo ergaster* into a single chronospecies, mainly on the grounds that little had changed between the two other than a further modest increase in brain size among the later Asian specimens.

The big story here, in many ways, is the first exodus out of Africa (Fig. 5.1). We know that this occurred very early in the history of *ergaster*: they were in Georgia, north of the Black Sea, by around 1.8 million years ago, since fossils of a derivative – if rather smaller-brained – species, *Homo (erectus) georgicus*, have been found there dated to this period. If the dating for these fossils is correct, then it implies that *Homo ergaster* invaded Eurasia quite soon after it first appeared in Africa – southern Russia is not exactly next door and must have taken some time to reach – and did so repeatedly in waves that resulted in ever larger-brained populations colonizing Eurasia.

The quite modest differences in material culture and brain volume over the time range of this chronospecies – the cranial capacity got only slightly larger with time, but handaxes changed hardly at all in design over the course of a million years[1] – mean that the main thing we really have to account for is the rapid initial increase in brain volume that occurred when the taxon first appeared in Africa around 1.8 million years ago (Fig. 1.3) and then the additional modest increase that defines Asian *erectus*. Any growth in brain size will only be possible if the species increases its food intake to provide the extra energy needed for it. Given that australopithecine time budgets were already at their limit with no spare

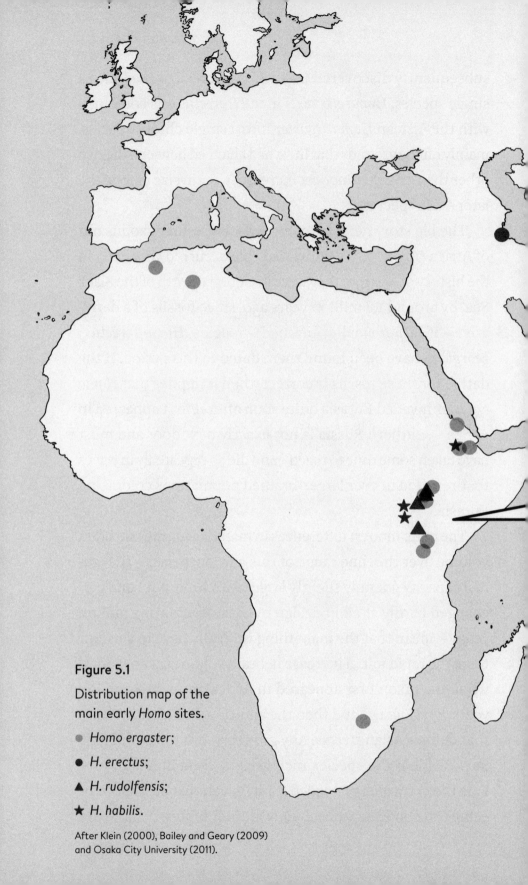

Figure 5.1

Distribution map of the
main early *Homo* sites.

- *Homo ergaster*;
- *H. erectus*;
- ▲ *H. rudolfensis*;
- ★ *H. habilis*.

After Klein (2000), Bailey and Geary (2009)
and Osaka City University (2011).

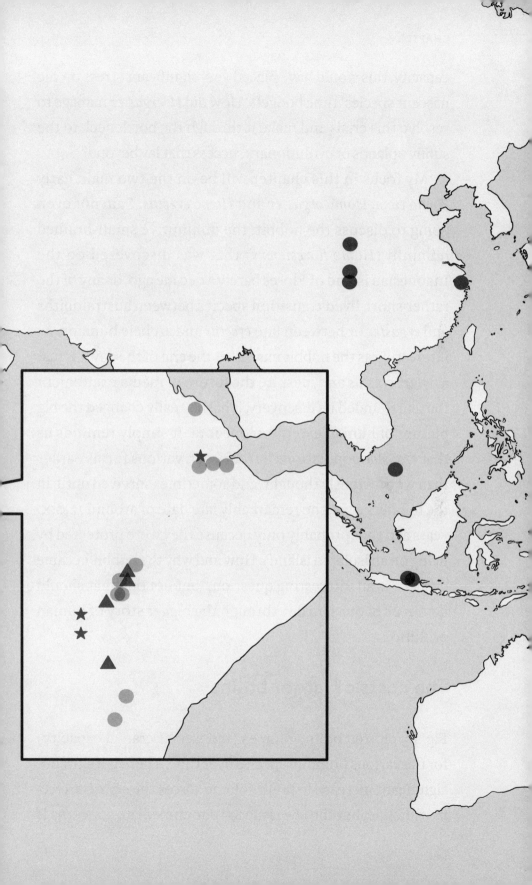

capacity, this would have placed very significant stress on the nascent species' time budgets. How did *H. ergaster* manage to resolve this crisis and make it through the bottleneck to the sunny uplands of evolutionary success that lay beyond?

My focus in this chapter will be on the two main early *Homo* taxa, *Homo ergaster* and *Homo erectus*. I am not even going to discuss the hobbit, the diminutive small-brained hominin (*Homo floresiensis*) that was discovered on the Indonesian island of Flores barely a decade ago, or any of the rather short-lived transition species between australopiths and *ergaster* or between late *erectus* and archaic humans. As interesting as the hobbit may be, at the end of the day it is just a dwarf *erectus* and, despite the furore of media excitement that surrounded its discovery, it hasn't really changed the big picture of human evolution one iota. It simply reminds us that *ergaster* populations left Africa in various forms earlier than we previously thought, and sometimes survived until, in the hobbit's case, the remarkably late date of around 12,000 years ago (but probably only because they were protected by living on an isolated island). How and why the hobbit became dwarfed is an interesting question,[2] but not one that should detain us in our journey through the bigger story of human evolution.

The costs of larger brains

Fig. 1.3 plotted brain volumes (indexed as cranial capacity) for the various hominin species over time. There is a slow but significant increase in brain volume across the *ergaster*/*erectus* timeline, but the key issue for our immediate concerns is

that brain volume increased from around 480 cc in the australopithecines (a fraction above that for living chimpanzees) to
an average of around 760 cc for the five earliest specimens of
H. ergaster in East Africa. This represents an increase of 280
cc, or 58.2 per cent, over the australopithecines. There is a
further slight increase over *ergaster* of about 170 cc to 930 cc
(22 per cent) in later *Homo erectus*.

This growth in brain size would have required significant
energy to maintain, which in turn means more foraging time.
We can estimate exactly how much extra foraging time would
have been required by doing some very simple calculations.
The adult human brain of 1,250 cc consumes about 20 per cent
of the body's total energy turnover, so an increase of 280 cc
represents 280/1,250 = 22.4 per cent of the brain's energy cost.
Since this represents 20 per cent of the total energy cost for
the whole body, this 22.4 per cent extra brain energy requirement represents 22.4 × 0.2 = 4.5 per cent extra energy intake
overall. If we assume that energy requirement directly corresponds to time spent foraging, on average, then the extra
energy needed to fuel *ergaster*'s brain represents an increase
of 4.5 percentage points on the australopithecines' energy
intake. The additional increase in the size of *erectus*'s brain
would have added another 2.7 percentage points, raising their
feeding time by a total of 7 percentage points over that of the
australopithecines.

Caroline Bettridge's models suggested that australopithecines satisfied their gross energy demand by devoting 44 per
cent of their day to feeding (Fig. 4.3). Assuming that energy
intake corresponds more or less to time spent feeding (it
will vary with diets, but broadly speaking this is a reasonable

assumption), the extra brain volume means that *ergaster* would have needed to spend 44 + 4.5 = 49 per cent of its day feeding, and *erectus* would have had to raise that to about 51 per cent. An increase of about 5 percentage points is not excessive, and could probably have been absorbed by small adjustments in the other activity categories. The slightly higher feeding demand for *erectus* would, however, have been more difficult to massage away. On balance, then, it seems that these early increases in brain size would have imposed quite modest costs on early *Homo* – rather more modest costs, in fact, than most human evolution researchers have supposed. Nonetheless, as with the australopithecines and their need to find a 7 percentage point overspend on their time budget, so *ergaster* and *erectus* would have needed to find savings of about 5 and 7 percentage points to allow their increase in brain size.

Brain size was not, however, the only aspect of anatomy to have changed with the advent of *Homo*. The other major change was a combination of body size and shape: *H. ergaster* was taller, with longer legs designed for striding (or, some have suggested, jogging in pursuit of prey) and they were quite a bit heavier. Averaging across all species, male and female australopithecines weighed about 55 kg and 30 kg respectively, while the equivalent figures for *Homo ergaster/erectus* were 68 kg and 51 kg. On average, early *Homo* was 40 per cent bigger than the average australopithecine and this extra body mass would have required its own additional energy. Taking these two components (brain and body) together, *ergaster* would have needed to spend 62.5 per cent of its time feeding in order to meet its somatic (or bodily) and cerebral energy demands – an increase of 21.5

percentage points above that for the australopiths.[3] The figure is almost identical for *erectus*, because although its brain was larger than *ergaster*'s, there was a compensating decrease in body size – possibly a consequence of adapting to a more stressful, colder, more seasonal environment at higher latitudes in Eurasia. Most researchers have assumed that it is the extra costs of large brains that was the problem that *Homo* faced, but these calculations suggest that, in fact, the real cost was the relatively much larger increase in body size: this accounts for more than three-quarters of the total extra energy (and time) demand.

An additional 22 percentage points for feeding would have made the standard australopithecine activity budget – already at its limit, as we saw in Chapter 4 – completely untenable. Fig. 5.2 illustrates the scale of the problem. This figure plots the percentage of total day time that each of the main hominin species would have had to devote to feeding (*not* foraging, as the travel time component of foraging is calculated separately), taking into account just the costs of increased brain and body size. The australopithecine reference line (44 per cent of time, as estimated from the Bettridge model) is indicated by the horizontal line. The various australopith specimens, including *Homo* (*Australopithecus*) *habilis*, all cluster comfortably around this line. But the appearance of *Homo ergaster* marks the start of a dizzying increase that rises to 62 per cent for early and archaic *Homo*, 78 per cent for Neanderthals (reflecting the large size of both their bodies and their brains) and 72 per cent in anatomically modern humans (reflecting a slight decrease in body size compared with the Neanderthals).

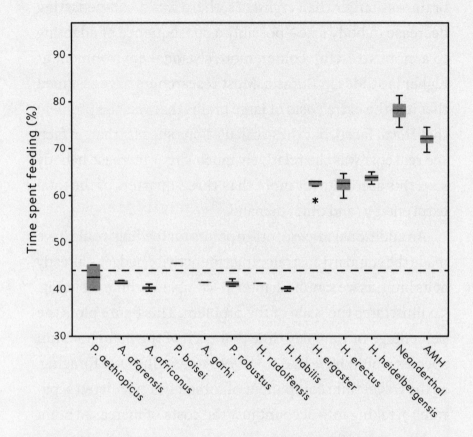

Figure 5.2

Median (with 50 and 95 per cent range) feeding time for the main hominin species, calculated for each specimen by scaling the mean australopithecine time budget (reference line at 44 per cent) as a function of the ratio of the species' metabolic body mass relative to that for australopithecines. Metabolic body mass is allocated differentially between the energetically more expensive brain (accounting for 20 per cent of gross energy) and body mass (accounting for 80 per cent).

This was not the only problem facing *ergaster*: it also had to cope with the additional social time required to bond the larger community sizes implied by its bigger brains. Fig. 5.3a plots the social time requirement for the hominin species (using the relationship between group size and grooming time shown in Fig. 2.1), and Fig. 5.3b plots the combined time budget, including travel and rest time at the fixed rates predicted by the australopithecine model (16 per cent and 32 per cent respectively). It will be apparent that early *Homo* will have been pushed far above its sustainable time budget (they would have been in deficit by about 30 per cent). And the situation simply gets worse for archaic humans: both Neanderthals and anatomically modern humans would have been a completely unsustainable 50 per cent over budget.

In short, the net savings that the *ergaster* populations would have had to find were a very significant 30 percentage points, two-thirds of which would have been for extra feeding time. *H. erectus* would have been 34 percentage points over budget, mainly due to the fact that its larger brain implies a further increase in group size. The bottom line is that none of these species could actually have existed had they been australopithecines. Somehow, they must have found ways to circumvent the time budget bottleneck. So how did they do it?

Some possible solutions

Resting might seem an unlikely option because the australopithecines had probably exhausted what little capacity there was in this respect through the advantages of bipedalism and hairlessness. However, there is one way in which

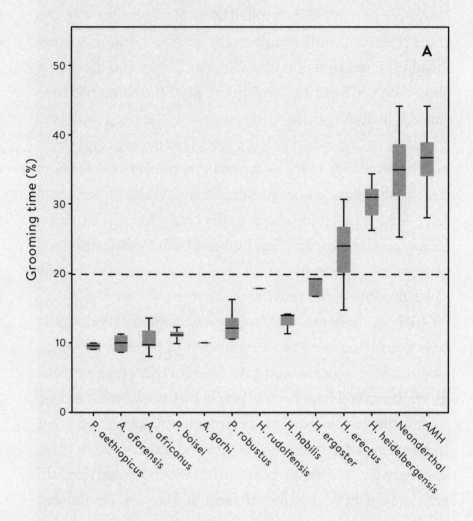

Figure 5.3

(a) Median (with 50 and 95 per cent range) social (grooming) time for the main hominin species, calculated for each specimen by interpolating community size from Fig. 3.3 into the regression equation for Fig. 2.1. Dashed line at 20 per cent indicates maximum time devoted to social grooming by free-ranging primates.

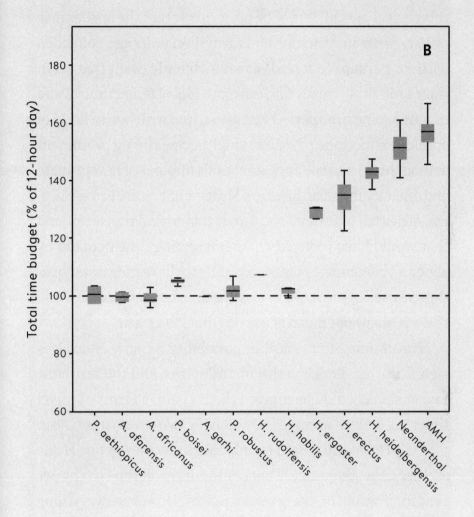

(b) Median (with 50 and 95 per cent range) total time budget for the main hominin species, calculated by adding feeding time (Fig. 5.2) and social time (Fig. 5.3a), with constant enforced rest time (32.5 per cent) and travel time (16.2 per cent) from the Bettridge (2010) australopithecine model (see Fig. 4.3).

ergaster might have been able to gain some additional savings on resting time, and that is if the climate was cooling during this critical period. It may be no accident that the appearance of *H. ergaster* in Africa around 1.8 million years ago coincided with (or perhaps followed) a major climatic event that set in train a relatively steady fall in average global temperatures and that this same time period was associated with both a flood of species extinctions (the australopithecines being prominent among these) and the appearance of a number of new primate and other mammalian lineages. Mean temperature across tropical Africa fell by around 2°C across this critical transition, and that would have reduced *ergaster*'s resting time demand by about 2.5 percentage points according to the resting time equation we used in the monkey and ape time budget models. Well, that's perhaps not much of a saving, but it's a start.

Travel time offers another possibility, as early *Homo* was significantly taller than the australopiths, and the resultant longer stride would have given them a gain in terms of travel time. Leg length (averaged between the two sexes) was 0.88 m for early *Homo* and 0.62 m for australopithecines, giving *Homo* a 41 per cent advantage. That could cut travel time to 11.5 per cent for a saving of 5 percentage points. However, everything about early *Homo* (including its long legs) suggests that it was much more nomadic than its predecessors and may therefore have ranged much more widely on the savannahs. It may be that this particular saving had more to do with facilitating longer day journeys for the same time cost than with effecting savings in travel time. However, for the moment, let's consider it as a small possible saving, giving us a gross saving so far of 7.5 percentage points.

Since feeding remains the dominant element in early *Homo* time budgets (it accounts for nearly 62 per cent of total time for both taxa), savings on feeding time are likely to have had a bigger effect. One possibility that we haven't considered is that there was some reorganization of energy allocation so as to redistribute the costs of somatic maintenance between different body parts. This is what Peter Wheeler and Leslie Aiello had in mind when they proposed their expensive tissue hypothesis in 1995. They noticed that the gut and the brain are equally expensive in terms of energy consumption, mainly because guts are highly innervated: it is the cost of maintaining neurons in readiness to fire that is the expensive part of the brain. They suggested that at some point in human evolution hominins switched energy allocation from one expensive form of tissue (the gut) to another (the brain) so as to effect an increase in brain size at no extra cost, and that they did this by improving diet quality so as to improve the rate of nutrient extraction to compensate for the reduced gut size.[4] In making the calculations of hominin time budgets for Fig. 5.3b I made no allowance for this but instead assumed that body mass was distributed among the main organs just as it had been in australopithecines, aside from the relative increase in brain mass. But what if the extra brain cost had in fact been traded for an equivalent reduction in the cost of the gut? How much would this have saved them in terms of feeding time?

If we recalculate hominin feeding time requirements leaving out the brain size component (i.e. by adjusting only for the increases in body size, assuming that the increase in brain size was directly compensated by a proportionate decrease in gut size), then we get modest savings of about 5 percentage points in early *Homo* and slightly larger savings in later hominins (Fig. 5.4). Though not as large a saving as most researchers would probably have anticipated, a 5-point saving would bring the time budget overspend for *ergaster* and *erectus* down from 30 and 34 percentage points, respectively, to 25 and 29 percentage points – of which, we have so far managed to account for 7.5 percentage points through climate cooling and leg length. That leaves us with 17.5 percentage points to account for in the case of *ergaster*, and 21.5 percentage points in the case of *erectus*.

What other options might they have had? Meat has been one suggestion, on the grounds that meat is a significantly richer food source than most plant foods, as well as being packed full of key micronutrients like the vitamin niacin (B_3), which is crucial for brain development. This was the basis of the original expensive tissue hypothesis: the reduction in gut size was traded off against a more digestible diet, and this more digestible diet was assumed to be meat. However, the meat option isn't quite as simple a solution as was originally assumed: raw red meat is not easily digested by primates. The Harvard anthropologist Richard Wrangham has argued very cogently that cooking might have solved the energy gap in human evolution by increasing the digestibility of meat: cooking increases nutrient extraction rates from meat by about 50 per cent. His favoured date for the invention of cooking (and hence meat-eating) is the appearance of *Homo ergaster* around 1.8 million years ago. He maintains that this suggestion is bolstered by three key archaeological observations: the appearance of the genus *Homo* coincides with a

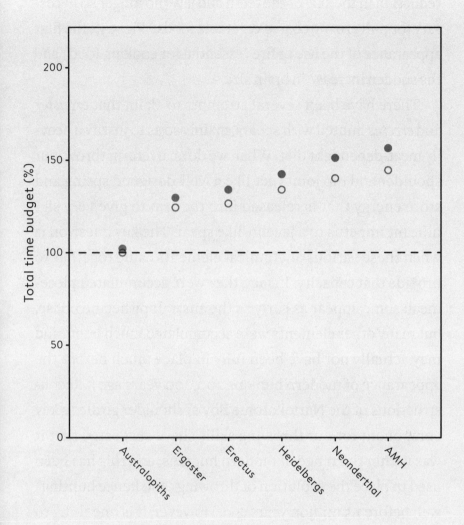

Figure 5.4

Savings on total time budget due to the expensive tissue effect on feeding time. Baseline time budget (species values scaled up from the australopithecine baseline shown in Fig. 5.3b), adjusted to allow for the expensive tissue effect. The expensive tissue adjusted value assumes that the additional energetic cost of a larger brain is directly offset by a corresponding reduction in gut (somatic tissue) mass; hence only body mass (at the reduced metabolic costs of somatic tissue) influences feeding time.

reduction in the size of the teeth and jaw (no longer so necessary for pulverizing coarse vegetable foods:[5] Fig. 5.5), the first
appearance of the use of fire (essential for cooking food) and
the sudden increase in brain size.

There have been several attempts to claim that *ergaster*
and *erectus* hunted with spears, mainly so as to justify a heavily meat-dependent diet. When we do an overarm throw, our
shoulder and hip joints act like a well-designed spring and
store energy that is released into the arm to give very significant impetus to a javelin-like spear. The key question is
when these various anatomical elements came together to
provide that capacity. In fact, they were accumulated piece-
meal: some appear as early as the australopithecine phase,
but many other elements were accumulated much later (and
may actually not have been fully in place much before the
appearance of modern humans 200,000 years ago). Reconstructions of the Nariokotome Boy's[6] shoulder girdle (a key
component for our throwing skills) have suggested that it
was within the range of modern humans, and this has been
used to place the evolution of throwing (and hence hunting)
well before 1.5 million years ago. However, it is one thing to
assert that the *erectus* shoulder girdle was in the low range
for modern humans and an entirely different matter to claim
that the species' throwing abilities were as good as those of
modern humans. After all, not all of us are Olympic javelin
champions or major league baseball pitchers, and exactly
where we lie in the range makes a great deal of difference to
how well we throw (not to mention how many gold medals
we get). On balance, the claim that early *Homo* hunted with

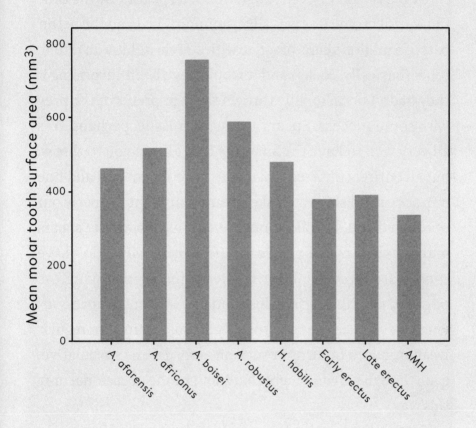

Figure 5.5

Mean molar tooth area for early hominins and modern humans.

Source: Leonard et al. (2003)

spears on a regular basis seems to stretch the available evidence beyond the bounds of plausibility.

On the other hand, it is quite another matter to suggest that *erectus* might have obtained the occasional prey by throwing rocks (or even rough wooden spears). By the time the Nariokotome Boy was alive, hominins had had the better part of a million years' practice with simple (Olduwan) stone tools (basically, rocks) and it would have been surprising if they hadn't occasionally thrown them at predators or prey with some success. Spears are another issue, perhaps. It is all very well to have the anatomy that allows you to throw, but an entirely different matter to have the mental and fine motor control skills to make spears that flight properly out of rough wood. It is also questionable as to whether spears without good blades (and early *Homo* certainly didn't have these) are effective hunting weapons for throwing (as opposed to stabbing). Unless and until we have more convincing evidence, we can probably dismiss claims for regular weapon-based hunting at such an early date as speculative, even though we can accept that they got the occasional prey this way.

Setting the hunting issue aside, however, cooking would certainly increase the rate at which nutrients can be extracted from foods like meat. It does so partly by breaking down the cell walls (thus aiding nutrient digestion) and partly, in the case of some plant foods, by detoxifying the alkaloids and other toxins that the plants include in their vegetative parts to reduce browsing damage by herbivores. However, cooking does not necessarily improve the digestibility of all foods equally: the real benefit is gained only from red meat and underground storage organs, whose rich but indigestible starches are softened by the cooking process. However, even contemporary hunter-gatherers do not eat an all-meat diet: meat and underground storage organs between them typically account for only about 45 per cent of all foods eaten (Table 5.1).

So let's see what cooking might have done for early *Homo*. If, taking the modern human pattern as a benchmark, 45 per cent of their diet benefits from a 50 per cent increase in digestibility, then the overall quality of the diet would be improved by 22.5 per cent.[7] In effect, every 100 percentage points of feeding time buy the equivalent 122.5 percentage points in terms of nutrient intake. Since *ergaster* and *erectus* needed to devote 57 per cent and 58.5 per cent of their time, respectively, to feeding to fuel their bigger bodies and brains, cooking would have reduced their feeding time demand to $57 \times (100 \quad 122.5) = 46.5$ per cent for *ergaster* and $58.5 \times (100 \quad 122.5) = 47.8$ per cent for *erectus* (savings of 12 and 9 percentage points, respectively). In effect, cooking the whole of their diet would have had a modest impact on their time budget, reducing their overall deficit, but it would still have left us, respectively, about 5.5 and 13.5 percentage points short. They could, of course, have met this remaining deficit by increasing the proportion of meat or underground storage organs in their diet. For *ergaster*, that would have meant increasing the meat and/or root component by 50 per cent up to 67.5 per cent of the diet, while for *erectus* it would have meant an increase of 130 per cent to 105 per cent of the diet. Both figures seem high given the limited archaeological record for meat during this phase, but the

FOOD SOURCE (% OF DIET)	WORLDWIDE	TROPICS	TEMPERATE
Plant foods	26.7	40	40
Tubers, roots, bulbs	8.3		
Meat	36	34	40
Fish	29	26	20

Table 5.1

Diets of modern hunter-gatherers.

Source: A sample of 63 modern hunter-gatherer societies from the Ethnographic Atlas (Cordain et al. 2000).

erectus figure is clearly impossible: they would have had to be totally dependent on meat (or some combination of meat and underground storage organs) – something that even modern humans don't do.

Wrangham's second piece of evidence is that molar tooth size is reduced in early *Homo*. It is undoubtedly true that *Homo ergaster* molars are a lot smaller than those of the robust australopithecines, but they are not in fact *that* much smaller than those of the gracile australopiths (Fig. 5.5). The impression of smaller teeth in *Homo* has, I suspect, largely been due to the fact that such data are invariably presented relative to body mass. Unfortunately, there is a widespread tendency to automatically relate everything to body mass in the palaeoanthropological literature, but this is misplaced unless there are good *biological* reasons for doing so. This is one of those cases: it is not relative molar size that determines the chewing area available, and hence chewing efficiency, but *absolute* molar size. The gain is thus much smaller than has been assumed. The third piece of evidence is the increase in brain size that occurs with the appearance of *Homo*. But in reality, the increase in brain size at this point in hominin history is quite modest compared to what came later (Fig. 1.3). More importantly, if we use cooking to explain this small brain size increase in early *Homo*, what will be left to explain the much larger increase in brain size that comes with archaic and anatomically modern humans?

So whatever cooking was doing, it almost certainly wasn't providing the complete solution. Indeed, it is not even clear that the modest increase in brain size that occurs with the appearance of *Homo* couldn't have been solved without needing to resort to a lot of meat and without needing to cook it. After all, meat isn't *totally* indigestible raw; you just don't get quite as much out of it as you do by cooking it. Chimpanzees eat modest amounts of meat – and relish it – despite not cooking it, and there is no reason why early *Homo* might not have increased its meat intake somewhat even without cooking. The question is simply whether it would have mattered that they were losing a third of the possible nutrients available when they did eat meat. Adding 5 percentage points extra from meat-eating to pay for the expensive tissue effect translates into a rather modest increase from something less than 2 percentage points of total feeding time in chimpanzees to something like 5–7 per cent in early *Homo*. Cooking would have reduced that to something like 3.5–4.5 percentage points, for a very modest saving – so modest, in fact, that I am not even sure it would have been worth the effort of lighting the fire.

What is, I think, clear is that early *Homo* were unlikely to have relied heavily on meat to get them off their time budget hook. However, the real issue is that, if Wrangham is right and cooked meat (or roots) was the solution of choice for them, then we ought to be seeing lots of archaeological evidence for fire, because without fire there can be no cooking. So, what does the archaeological record tell us about the use of fire?

The evidence for fire

Fire provides both heat and light, and was unquestionably crucial in allowing humans to invade high latitudes in Europe and Asia (and, of course, eventually the Americas) as successfully as they were later to do. Our main concern for now, however, is whether fire might have evolved early enough that *Homo ergaster/erectus* could have used it for cooking. If fire is to account for even as much as half of the diet, then it must surely have been in more or less continuous use, and we should expect to see widespread evidence for hearths.

The oldest sites that show convincing evidence of fire use by hominins (indicated by burned soil) are at Koobi Fora and Chesowanja in Kenya, both dated to around 1.6 million years ago – so quite soon after the first appearance of *Homo ergaster*. However, after this, there is nothing until around 1 million years ago, when once again we have sporadic evidence for fire. The earliest case from this second phase is some 60 pieces of charred bone at Swartkrans in South Africa. The fact that the bones have been subjected to relatively high temperatures, combined with some evidence of cut marks, has been used to argue that this was deliberate cooking rather than bones accidentally falling into fires. Burned bone recovered from Bogatyri on the north side of the Black Sea has been dated to 1.1 million years ago, while the site of Gesher Benot Ya'aqov in Israel (dated to around 0.7 million years ago) has yielded burned wood and seeds. At Wonderwerk, a South African site of roughly similar age, evidence of burning is associated with Acheulean artefacts: nearly half of the bone fragments at this site show evidence of discoloration due to burning, and there are a number of ashed plant remains. Then, after this, there is yet another long period devoid of evidence for hearths.[8]

In contrast, evidence for fire use from 0.5 million years ago onwards is abundant from sites widely distributed across all three Old World continents. Hearths have been found at

Tabun, Qesem and Umm Qatafa in Israel, Beeches Pit in Britain and Schöningen in Germany (all dated to around 400,000 years ago). At Schöningen, there is a charred wooden stave, while at Clacton (a site of similar age in England) a wooden spear had clearly been fire-hardened. At Beeches Pit in Suffolk, two flakes from a handaxe had fallen into the hearth and been burned: here, caught in a frozen moment in time, was one individual making a flint axe by a hearth. Uncontroversial hearths have been found at Abric Romani in Spain, Terra Amata in the south of France, the Menez Dregan rock shelter in Brittany and Grotte du Lazaret on the French Mediterranean coast, all dating after 400,000 years ago. In fact, almost everywhere you look after 400,000 years ago there is evidence of hearths.

A hearth where Neanderthals cooked small mammals (mainly rabbits) has been reported from Bolomor in Spain. In Africa, Kalambo Falls in Zambia (dated to 180–300,000 years ago) has yielded charred logs as well as burned digging sticks and a club, and hearths have been found at Florisbad (dating to the last interglacial) and at Pinnacle Point (dated to 40–170,000 years ago) in South Africa. After 100,000 years ago, there are many examples of hearths at sites such as La Roca dels Bous in Spain, the many Mousterian sites in Moldova (including the famous camp of the mammoth-bone huts) and the interglacial German site at Wallertheim.

The scarcity of hearths in Africa, Europe and Asia prior to 0.4–0.5 million years ago has been interpreted as suggesting that this date marks a major step-change in the use of fire. Before this point, hominins might well have exploited natural fires (such as those from lightning strikes) casually and opportunistically, but did not have sufficient control over fire to keep hearths alight. It seems that fire was fully mastered around 400,000 years ago, and once mastered could be maintained and rekindled anywhere and at will. This step-change in fire use seems to coincide with the appearance of regularly used home bases (including caves as well as huts). Large hearths probably require in excess of 30 kg of wood a day, and are very demanding on time, energy and coordination since someone has to be responsible for collecting all this firewood. Doing this on a daily basis would add significantly to already overstretched time budgets. More importantly, it may only be possible to keep large fires burning if several people coordinate their activities – something that probably requires language as well as the cognitive ability to recognize the need for cooperation and turn-taking. Both of these are likely to be dependent on a large brain: indeed, I shall argue in Chapter 8 that brains large enough to support these kinds of cognition are unlikely to predate the appearance of archaic humans around 500,000 years ago. In sum, although there is certainly putative evidence of cooking much earlier – usually in the form of charred bones or seeds – the evidence strongly suggests that cooking did not become a *regular* feature of the diet until after 400,000 years ago. If this interpretation is correct, then we probably have to conclude that cooking was not regular enough to have had a significant impact on the diet of early *Homo* and, hence, could not have been the main solution to their time-budgeting crisis. As with many other phenomena in human evolution, much of the literature on fire has focused on finding the earliest possible case of its use. But it is not enough to demonstrate that fires or cooking occurred once; rather, to be able to argue that cooking played a crucial role, we need to be confident that fire use occurred *regularly*, day in and day out, week after week. So while casual cooking of meat may well have occurred on quite a regular basis, it didn't do so on such a regular basis that it would have made a significant dent in the early *Homo* time budget deficit.

How laughter solves the bonding problem

The one time budget component that we haven't considered yet is social time. The social time requirement directly relates to community size, and the much larger community sizes of early *Homo* would have virtually doubled the grooming time requirement for these taxa compared to the australopiths. Our assumption in calculating the social time requirement has been that groups are bonded through social grooming, just as in the monkeys and apes. In primates, grooming is very much a one-on-one activity: it is physically impossible to groom several individuals at the same time. Indeed, that problem is still with us: the equivalent for modern humans is the kind of physical affection that we share in more intimate moments, and it isn't something you can do with several people at once (at least, not without one of them getting upset). The real issue, however, is that the amount of time that monkeys and apes spend in social grooming seems to reach a limit at about 20 per cent of the day, mainly because of the demands of the other time budget components. And that, in turn, sets an upper limit on the size of group that can be bonded at about 50 individuals. When early *Homo* needed to increase community size above this limit, either they had to make room for more grooming time *or* they had to find a more efficient way of bonding that allowed more people to be bonded in the same amount of time. In effect, they needed to be able to groom with several individuals at the same time.

Primates, as we saw in Chapter 2, rely on the physical stimulation provided by social grooming to trigger endorphin activation in the brain, and it is this endorphin activation that appears to be critical for bonding. As it happens, there is one behaviour that might just allow several people to be 'groomed' simultaneously: laughter. Humans share laughter with the great apes. Ape laughter – which is normally used in a play context – is a series of exhalation–inhalation sequences, whereas in humans laughter consists of a series of exhalations without accompanying inhalations. Because apes inhale after every exhalation, they do not empty the lungs, and the pressure on the diaphragm and chest wall muscles is minimal. In contrast, the rapid series of exhalations in human laughter empties the lungs, leaving us exhausted and gasping for breath. Since, as we have shown in a series of experiments, this stress on the chest muscles triggers endorphin activation, then it might be possible to use laughter as a form of grooming at a distance, and so trigger an endorphin effect in several individuals at the same time.

Laughter is the perfect candidate for this because not only does it trigger endorphin activation in the brain, but few human behaviours are quite as contagious. Laughter is up to 30 times more likely to occur when watching a comedy video in a group than when watching the same video alone. Indeed, so instinctive is this behaviour that we have difficulty not laughing when everyone else is laughing, even though we haven't understood the joke. In short, laughter has all the hallmarks of social chorusing. The fact that this seems to be instinctive suggests that it has a very ancient origin; the fact that we don't need verbal jokes to trigger laughter suggests that it long predates the evolution of language.

The critical issue for our analysis is how many more people can be influenced by laughter than by grooming. In other words, what is the typical number of people involved in a spontaneous bout of social laughter? To determine this, Guillaume Dezecache carried out a survey of social laughter in pubs. He recorded the size of social groups (essentially the number of people sitting round a table who were obviously together as a group), the size of conversation groups (the number of people within a social group that were actively involved in the same conversation, either as speaker or paying attention to the speaker) and, when laughter occurred, the number of people actively laughing at the same time (the laughter group). In this particular sample, the average size of social groups was about seven, but conversation and laughter groups were *much* smaller than this. As we had found in a previous study, conversation groups had an upper limit at four: if more than four people get involved in a conversation, the group will very quickly split into two separate conversations. The real surprise was that laughter groups (i.e. the set of individuals laughing *with* each other in a social context) were even smaller: they had an upper limit at around three. This was much smaller than we had anticipated, especially given that so much of our laughter is now triggered by jokes using language.

Because all three members of the laughter group (the one who tells the joke *and* the ones who hear it) laugh, all three experience an endorphin surge. In grooming, it is only the *recipient* of grooming that gets the hit. In effect, laughter is three times more efficient than grooming as a bonding process. If hominins adopted laughter as a bonding mechanism in addition to grooming at this early *Homo* stage, it would have allowed them to save a significant amount of time. Our equation for social grooming as a function of group size in Old World primates tells us that, with groups of around 75, *ergaster* would have needed to devote 18.5 per cent of their day to social interaction, while *erectus*, with groups of roughly 95, would have needed 23.5 per cent. If laughter supplemented grooming and was three times more efficient than grooming on its own, then it would have reduced this requirement to a third, or 6.2 per cent and 7.8 per cent of the day, respectively. This would bring us savings of 12 and 15.5 percentage points for the two species, which, combined with the 12.5 percentage points we gained from climate change, bipedalism and the expensive tissue hypothesis, would allow us to save 24.5 and 28 percentage points from the two species' time budgets.

In sum, this leaves us just 5–6 percentage points short of the 30 and 34 percentage points that we need to balance the two species' time budgets even without having to invoke cooking (Fig. 5.6). This relatively small outstanding balance could easily be accounted for by a very modest amount of casual cooking: it would require only 5 per cent of the diet to be cooked, making early *Homo* much less dependent both on maintaining hearths and on killing prey, but at the same

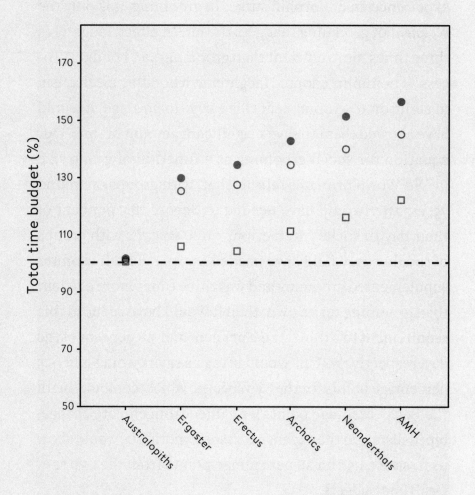

Figure 5.6

The impact of laughter on total time budgets in the main hominin species. The baseline total time budget (●) and the expensive tissue adjusted total budget (○) are as in Fig. 5.4; the effects of savings in the social time requirement due to laughter are shown as open squares (□).

time setting them up with a capacity that later species might be able to expand.

I have taken the estimate for *Homo erectus* community sizes based on its larger brain size at face value for the moment. In Chapter 6, I shall show that Neanderthals had smaller community sizes than one would predict for their cranial capacity because they had a much larger visual system as a result of living at high latitudes where light levels are significantly lower. All the *erectus* populations lived at similar latitudes to Neanderthals in Europe and Asia, and so must have faced the same selection pressure for improved vision and hence an enlarged occipital region (the back part of the brain responsible for primary visual processing). Although the evidence is limited, there is some indication that, like the Neanderthals, *erectus* had both larger eye sockets and a larger occipital lobe than their *ergaster* cousins. If so, then their frontal lobe may well have been similar to *ergaster*'s. If we correct *erectus* community size by the same proportion as we need to do for Neanderthals (see Chapter 6), the difference between the two species of early *Homo* pretty much vanishes – thereby neatly resolving the apparently higher time budget crisis faced by *H. erectus*. In other words, whatever solved *ergaster*'s time budget crisis also solved *erectus*'s.

Why did brain size increase in early *Homo*?

The increase in brain size that occurs with the appearance of *Homo* is significant and happened relatively quickly. What might have driven this dramatic increase? Perhaps the most widely accepted explanation across the hominins as a whole is known as the 'climate variability model'. This explanation notes that increases in brain size seem to occur at times when the climate is particularly unstable, involving rapid oscillations between wet and dry phases. Whenever the climate became especially unstable, hominins had to evolve a larger brain to cope with the more challenging foraging environment. Some versions of this hypothesis relate these cognitive changes to the production of better quality tools. On a more general note, the hypothesis is given some support by the fact that European birds that remain at high latitudes throughout the winter (and so face dramatic variations in food supply) have significantly larger brains than those species that migrate so as to overwinter in the tropics. Big brains certainly allow you to cope better with unpredictable environments.

The principal evidence for this hypothesis in hominins is that brain size correlates with climate fluctuations across the last 4 million years. However, this correlation seems to be the result of plotting individual specimens against geological time and ignoring the fact that they belong to different species. When we plot the same data for each species separately, the correlation vanishes, especially during the early *Homo* phase. The various changes in brain size are due entirely to species differences, not climate. Perhaps we shouldn't be surprised by this: there is no evidence that ecology affects brain size evolution in primates as a whole, and it would be odd if hominins suddenly, as if by magic, behaved completely differently from the rest of the family to which they belong. Nor is there any convincing evidence that tool complexity correlates with brain size. Tool complexity certainly increases over time, but the changes in tool complexity are out of step (and mostly postdate) changes in brain size. It seems that the climate variability hypothesis is not the answer, even though climate change may have been an important precipitating cause.

There is evidence to suggest that the dramatic increase in brain size that occurred with the appearance of early *Homo* coincided with a wet period when the East African rift valley at times consisted of a single large deep water lake running from the Omo valley in southern Ethiopia to Lake Baringo in central Kenya, no doubt with further large lakes further south. This may have greatly extended the range of some hominin populations, allowing them to occupy new habitats that had not previously been available. However, the ecological circumstances per se do not tell us *why* they needed big brains, but merely provide the circumstances that allowed them to be evolved. The fact that these rift valley mega-lakes frequently dried up (and often very rapidly) would, of course, have imposed very significant selection pressure on the populations that had occupied these new habitats: they would have had to adapt quickly to the more arid conditions or go extinct.

This leaves us with two possibilities. One is the old primate standard, predation risk; the other is threats provided by rival (hominin) communities – predators by another name. Predation risk is a plausible option for early *Homo*, because moving into more open terrestrial habitats must inevitably have resulted in an increase in exposure to predators. Baboons, for example, range widely out into the more open predator-risky woodlands and even grasslands beyond the forests they use as refuges, and they have the largest stable groups among the monkeys and apes; more importantly, these are invariably larger in populations that occupy more open savannah-type habitats. The extreme case is provided by the gelada, the grass-eating baboons of highland Ethiopia. This species lives in a complex fission–fusion social system based on small polygamous family units; the number of such units that band together in any given location depends on the predator-riskiness of the habitat. Their herds are largest when they forage on the open plateau tops where they have no trees in which to escape from predators.[9]

There is one problem with this suggestion, however. Both humans and chimpanzees face similar levels of risk from predators, yet there is no evidence to suggest that the higher (community) levels of social organization have anything to do with predation for either species. Because both species live in fission–fusion social systems, the members of these higher level units aren't physically close enough to act as useful deterrence to predators. The anti-predator grouping in chimpanzees is the foraging party of 3–5 individuals, and in human hunter-gatherers it is either the foraging group (perhaps 5 in the case of male hunting parties, and 10–15 in the case of women's gathering parties) or the overnight camp grouping (or band) of 30–50. The larger size of the overnight camps may reflect modern humans' vulnerability to predators at night, and hence the need to be in larger groups. Monkeys and apes always sleep in trees or on cliff faces, but humans cannot climb well enough to do this easily, and in general are obliged to sleep on the ground where they are extremely vulnerable to predators – and especially so given their primate-typical poor night vision. Either way, the higher level of organization in both species (the community of about 50 in the case of chimpanzees and 150 in the case of humans) is simply too geographically dispersed to function as a deterrence to predators.

Whatever function the community level of organization serves in chimpanzees, it is likely to be the same for early *Homo*, whose predicted community sizes are only slightly larger (roughly 75–80 in both taxa) than those of the chimpanzees. One plausible explanation for chimpanzee communities is defence of real estate, either to ensure the monopolization of resources or to defend reproductive access to the females that range within the territory. The first is possible, but would be unusual among primates: although many primates are territorial, in fact males invariably defend territories to monopolize access to reproductive females. The second would fit well with the factors that appear to select for territoriality in primates more generally and is offered significant support by the fact that male chimpanzees band together to attack the males of neighbouring communities – to the point of effectively wiping them out on occasion. Chimpanzee communities look suspiciously like self-defence alliances in which males gang together in order to protect both their own lives and their monopoly over the reproductive females that range within their defended area.

Although arguing that early *Homo* were just extending chimpanzee behaviour to larger areas and larger groups seems logical, in fact mate defence doesn't really make a lot of sense in their case. Why would they suddenly be obliged to extend the size of area they could collaborate in defending, necessitating a larger community size with its attendant brain-size costs? There doesn't seem to be any obvious reason for this. More importantly, it is clear that the delicate balance that allows a chimpanzee brotherhood to collaborate in defending a mating territory is not open-ended. In promiscuous mating systems, such as those of chimpanzees and baboons, the top-ranking male is unable to prevent other males mating with 'his' females once there are more than about five males in the group. His ability to monopolize females declines precipitously as the number of rival males increases, irrespective of the number of females in the group. Instead of being able to defend access to a group of females, the male is forced to focus on individual females as they come into oestrus. If more than one female is in oestrus at the same time, he just has to abandon her to another male. This is the pattern we see in both chimpanzees and baboons and macaques, so it has nothing to do with fission–fusion sociality, but is simply a consequence of the number of rivals with whom the dominant male has to contend. For a high-ranking male to be willing to tolerate more males in the group and share the supply of females with them there must be some pressing reason, and that reason has to be external to the group. If it is not predation, then it can only be either defence against raiding conspecifics (members of your own species), a problem that chimpanzees already experience, or the opportunity to share access to foraging resources over a wider area. Which of these explains the demand for larger communities, and hence larger brains, in early *Homo* remains unclear. I will return to this issue in Chapter 7, so for the moment I will leave the problem hanging.

One thing we can be sure of, though, is that early *Homo* was polygamous: the level of sexual dimorphism was a little lower than that in most of the australopithecines (about 1.25), but still higher than in modern humans, and indicative of a significant level of polygyny or promiscuity. Whether this reflects individual males competing for access to individual females as they became fertile (as in the case of chimpanzees and baboons), or that the females were divided up into harems monopolized by the more dominant males (as in gelada and hamadryas baboons, and perhaps the gorillas), cannot be determined with certainty. As with the australopiths, it probably depends on how large and how dispersed their foraging groups were. But given that community size was somewhat larger than in chimpanzees and australopiths, the most plausible conclusion is some form of polygamy.

So far, the story has been relatively straightforward. The changes in time budget involved in the transitions, first, into the australopithecines and, then, into early *Homo*, were modest and we have been able to deal with them fairly easily with relatively minor changes in diet, some gains from locomotion and, for the second phase at least, the introduction of a novel bonding mechanism (laughter as a form of chorusing). In large measure, this was because the increases in brain size and body size involved in these two transitions weren't as large as we are sometimes led to believe. The adaptations we introduced aren't insignificant – none of them was simply a matter of scaling up existing great ape traits – but, between them, they allow us to get both groups of early hominins 'on the road' without having to introduce cooking and fire. That is important because it means we still have cooking in reserve to allow for further changes in brain size to occur. This is just as well, because brain sizes get significantly larger and the story gets a great deal more complicated with the third transition into archaic humans.

The Third Transition

ARCHAIC HUMANS

Although the world's climate has never been stable over the long term, the last million years have witnessed rather more instability than had previously been the case. Measurement of oxygen isotopes from ice cores taken from the Greenland ice sheet as well as dust accumulations and oxygen isotope ratios in the shells of amyboid protists in deep sea cores off the Atlantic coast of Africa suggest that there have been a series of alternating cold and warm periods associated with the advance and retreat of massive ice sheets in the circumpolar regions.[1] At their peak, the ice sheets covered up to a third of the planet's land surface.

The temperature fluctuations associated with glaciations were quite substantial. Relative to current conditions, average temperatures were reduced by as much as 3°C in tropical latitudes, and by up to 16°C in Europe. With so much water locked up into the ice sheets, sea levels were up to 150 m lower than at present, thereby exposing submerged parts of the continental shelves and joining outlying islands like Britain to the neighbouring mainland. At their fullest extent, the ice sheets locked up as much as 50 million km³ of water, and greatly reduced the amount of atmospheric water available for rainfall: as a result, climates during glacial episodes were much drier than at present. Forests turned to grassland, and grassland into desert. Wind-blown dust levels in deep-sea cores increase sharply during glacial periods as a consequence of high levels of erosion on adjacent mainlands.

There have been eight full glacial–interglacial cycles in the last 750,000 years. Each glacial seems to involve a gradual build-up of ice (associated with declining temperatures) over a period of around 100,000 years, terminating with an abrupt warming that heralds the succeeding interglacial, most of which were quite short (lasting as little as 10,000 years in some cases). Glacial cycles are not actually characterized by continuous cold, but themselves involve smaller-scale oscillations between warmer and cooler periods.

The last full cycle (which includes the previous interglacial, the last glacial and the current warm, or interglacial, period) began 127,000 years ago. Around 115,000 years ago, the interglacial came to an end and temperatures began to cool steadily. In Europe, the forests gave way to woodland. Then from about 70,000 years ago, temperatures began a rapid decline and the last Ice Age set in with a vengeance. The last forest remnants disappeared from Europe and were replaced by open tundra characterized by tussock grasses and sedges that developed ahead of the advancing ice sheet. With the tundra came the herds of reindeer and saiga antelope, woolly rhino and steppe bison, musk ox and mammoths that had once dominated the landscape far to the north. Among the predators that moved southwards with them were giant cave bears, cave lions and cave hyenas, all of which were substantially larger than their counterparts of today.

By 40,000 years ago, the climate was fully glacial

throughout Europe and western Asia, reaching its greatest depth 18,000 years ago when much of northern and central Europe was locked in an icy grip. Then, 14,000 years ago, the climate began to take an upturn once more, culminating around 10,000 years ago in the end of the Younger Dryas Event[2] when temperatures rose by an astonishing 7°C in as little as half a century, precipitating the final end of the Ice Age. The ecological instability associated with these dramatic swings in temperature would have been very challenging.

The long haul of the *ergaster/erectus* clade[3] (over a million years from 1.8 to 0.5 million years ago with little change in either anatomy or material culture) is a reminder that when environmental conditions don't provide any selection pressure for change, none will occur. Then, quite suddenly around 600,000 years ago, a new species began to appear in Africa. Archaic humans (generally identified as *Homo heidelbergensis*[4]) had arrived.

It is clear that the *heidelbergs* managed to colonize Europe and western Asia at a surprisingly early date (they are recorded at Boxgrove around 500,000 years ago), and held pretty much continuous sway there over the next 200,000 years (Fig. 6.1). By about 300,000 years ago, however, the European *heidelbergs* had begun to evolve into early Neanderthals. Further east, they seem to have given rise to another species of archaic hominin, the Denisovans. So far, the Denisovans are only represented by a handful of bones from a single cave in the Altai Mountains of southern Siberia in deposits dated to around 40,000 years ago. Nonetheless, their DNA differs sufficiently from both the Neanderthals and anatomically modern humans to suggest that their Last

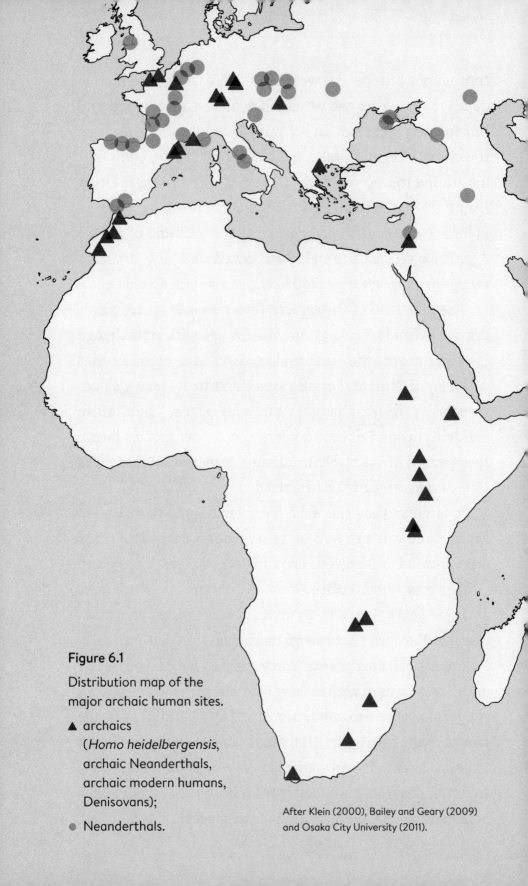

Figure 6.1

Distribution map of the major archaic human sites.

▲ archaics
(*Homo heidelbergensis*,
archaic Neanderthals,
archaic modern humans,
Denisovans);

● Neanderthals.

After Klein (2000), Bailey and Geary (2009)
and Osaka City University (2011).

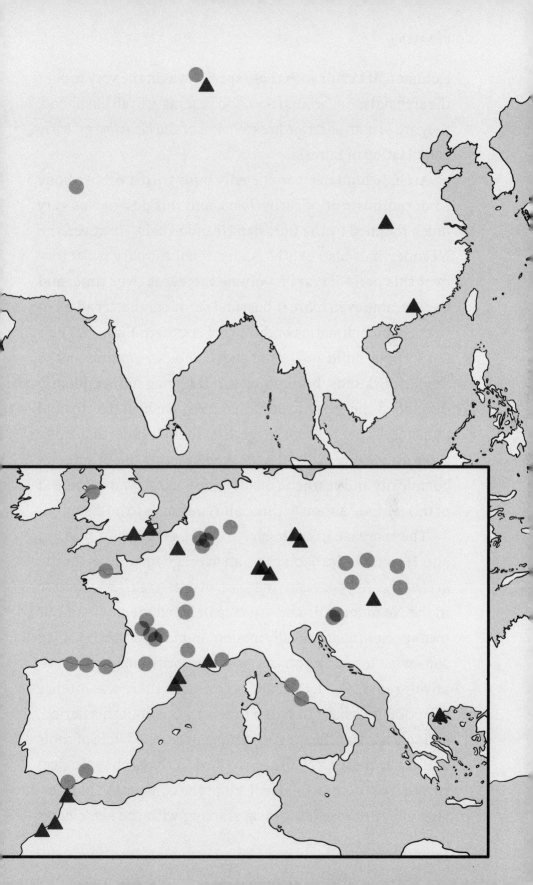

Common Ancestor with these species was at the very root of the archaic tree at around 800,000 years ago. In all likelihood, they are just an eastwards extension of the *heidelbergs'* early colonization of Eurasia.

Archaic humans were heavily built with a robust body form reminiscent of early *Homo*, and this design was very much retained by the Eurasian Neanderthals. However, the Neanderthals also exhibit a clear evolutionary trajectory over this period: cranial volume increases over time, and they became even more robust and seemingly better adapted to the cooler climatic conditions of southern Eurasia. Their stocky body build and rather short limbs are reminiscent of modern Eskimos, both populations having independently developed these as adaptations to reduce heat loss in cold climates. There is some suggestion that Neanderthal stone tools noticeably evolved over time in terms of both design complexity and aesthetic quality, especially towards the end of this period,[5] although this claim continues to be disputed.

The increase in brain size from *H. ergaster* (*c.* 760 cc) and *H. erectus* (*c.* 930 cc) to an average of 1,170 cc in the *heidelbergs* (a 55 per cent increase over *H. ergaster*), 1,320 cc in the Neanderthals and 1,370 cc in fossil members of our own species (anatomically modern humans, or AMH), all in somewhat less than 300,000 years, is both dramatic and relatively rapid (Fig. 1.3). This suggests that there was intense selection pressure for larger brains throughout this period, and this has significant consequences for the kinds of tools and other material culture we find in the archaeological record. For this reason, I will adopt the convention of referring only to these later taxa, starting with the *heidelbergs* and their allies, as humans. They seem to be in a genuinely different class to early *Homo*.

The first family

Sima de los Huesos (the 'Pit of Bones') in the cave complex in the Atapuerca Mountains, near Burgos in northern Spain, is one of the most remarkable hominin fossil sites anywhere. Discovered during the construction of the railway and briefly excavated in the 1900s, this site came to prominence during the 1960s and 1970s when major excavations revealed that the cave had been occupied by various human species over an unusually long time span. In 1983, excavators discovered the small, low-ceilinged chamber at the base of a 13 m chimney now known as the Pit of Bones. It has yielded up the largest collection of hominin fossils ever found: more than 1,000 bones from at least 32 individuals. Every bone in the body is represented, and the 32 individuals are a cross-section of ages and sexes. The site has been dated at 350,000 years ago, and it is arguable as to whether they were *heidelbergs* or archaic Neanderthals (as has been suggested more recently). Either way, it is a unique family-like collection of fossils that can stand for all archaic humans.

Like other archaic humans, the Pit folk show a mixture of modern and primitive traits. Their skulls are heavily beetle-browed, with low foreheads, and they lack an obvious chin. Yet brain size is a respectable 1,125–1,390 cc – on the small side by modern human standards, but significantly larger than that of any *ergaster/erectus* specimens. They are robust and heavy-boned. Their leg bones, for example, have a very thick cortical layer occupying almost the whole thickness of the bone (that of modern humans is barely a few millimetres thick), implying weight-bearing or other stresses during life. They were as tall as modern humans – on average, around 1.75 m for males and 1.70 m for females, with at least some individuals as tall as 1.8 m. In respect of their heavy brow ridge, sloping forehead, lack of chin and the distinctive gap behind the third molar, the Atapuerca people anticipate the later Neanderthals.

If the fossils represent a random sample of the population, then pre-adult mortality was clearly significant: half of all the specimens were under 18 years of age at death. Wear on the front teeth is particularly heavy, and associated (in nearly every individual) with arthritis in the bone surfaces where the jaw articulates with the skull: this suggests that the front teeth were often used as a vice for manipulating both food and non-food items (such as skins or tools). The growth lines in the teeth indicate that many individuals experienced physiologically stressful periods in their lives (associated with either illness or food shortage), many at around 4 years of age when weaning is likely to have occurred. On the other hand, their teeth are in remarkably good shape, with little evidence of dental caries. This perhaps reflects their frequent use of toothpicks – as evidenced by the presence of vertical grooves in the enamel around the gaps between the molars that look as though they have been caused by a fine object being used as a toothpick.

It is equally clear, however, that they suffered from high levels of illness. The Atapuerca 5 skull shows evidence of death from septicaemia (it seems to have begun in the teeth, and later spread all the way up into the eye socket and must have been *very* painful); another provides the earliest evidence of deafness (his ear canals were closed up by a kind of bony growth that is a typical after-effect of ear infections). There is no evidence of bone fractures, but several of the skulls have impact scars, possibly due either to tumbles or to being hit across the head with hard objects. Atapuerca 5 has an astonishing 13 such impact scars: he was either very careless or often got into fights.

Quite why the Atapuerca specimens ended up in the Pit is something of a mystery. The two most common reasons why bones accumulate in sink holes (carcasses taken back to cave lairs by predators, or individual bones being washed down by underground streams after an animal died at the surface) seem unlikely: at least some of the bodies are more complete than might be expected if they had been fed on by predators or individual bones had trickled down over long a period of time. Moreover, none of the bones shows any evidence of having been chewed. The possibility that the bodies might have been deliberately thrown down there after death – either to prevent their attracting predators, or to prevent their infecting the living area at the cave entrance as they decomposed – remains likely. Or perhaps the Pit was seen as the entrance to the underworld (a common theme of shamanic religions, as we shall see in Chapter 8) and represents a form of burial. Whatever the reason, disposal seems to have been relatively informal: the bodies lie higgledy-piggledy as though they had been tossed into the shaft.

Life in the Pit folk's world

In Chapter 5, we arrived at a feeding time requirement for *H. heidelbergensis* of 64 per cent (Fig. 5.2), just a fraction higher than that required by *H. ergaster*, which would have reduced to 56 per cent through the expensive tissue adjustment (Fig. 5.4). The *heidelbergs*' larger brains would, of course, have given them a significant increase in community size, so in addition to feeding time they would have needed extra social time to bond their larger communities. Our equations predict an average community size of about 125 for *heidelbergensis* populations, a significant 68 per cent increase above the 74.5 predicted for *H. ergaster* (Fig. 3.3). That would have necessitated a grooming time commitment of 30.5 per cent, almost 12 percentage points higher than *ergaster* needed (Fig. 5.3a). Their total time requirement, adjusted for the expensive tissue effect, would have been 134.5 per cent, almost 10 percentage points higher than *ergaster* (Fig. 5.3b). Of course, the continued cooling of the climate during this period might have reduced their resting time requirement, but only by a couple of points at most. They still had a significant overrun to contend with.

However, there is one issue that we have overlooked by focusing on species averages, namely the fact that *heidelberg* brain volumes seem to have undergone a rapid and quite dramatic increase around 300,000 years ago (Fig. 6.2). Although it is always difficult to be sure when sample sizes are small, the most plausible interpretation of Fig. 6.2 is that, after an early increase from its *ergaster* starting point, brain volume

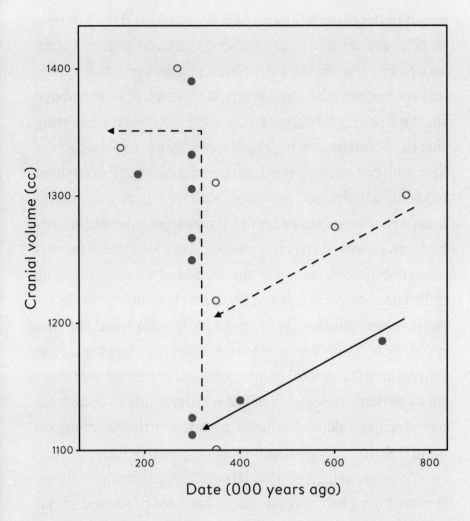

Figure 6.2

Cranial volumes of individual *Homo heidelbergensis* specimens over time.

- ● High latitude populations (Europe)
- ○ Low latitude populations (Africa)

The data appear to show declining brain size over time prior to 300,000 years ago as temperatures cooled, with an added effect due to latitude, followed by release from these constraints that may have been associated with the regular use of fire for cooking, warmth and, especially, lengthening the active day.

Source: De Miguel & Heneberg (2001)

began to decrease after 500,000 years ago and then underwent a very rapid increase sometime around 300,000 years ago. Notice that, in the early phase, the data points for tropical specimens (the open symbols) typically lie well above the data points for Europe (the solid symbols), suggesting that the occupation of high latitudes might have placed a very heavy burden on these northern populations and forced them to sacrifice brain (and, as we saw with *erectus*, possibly body) mass – presumably in order to balance their time and energy budgets. However, after the 300,000-year watershed, there is a very dramatic increase in brain size and, at the same time, the difference between high and low latitude populations disappears, suggesting that they might have found a solution to the problem. It is surely no accident that this date is just after the critical date for control of fire (400,000 years ago: see Chapter 5). Perhaps the control of fire and the ability to cook meat more regularly allowed archaics to overcome the constraint on brain size that they had been suffering prior to this.

Whatever triggered this late increase in brain volume, it would have had consequences both for predicted social group size and for total time budgets. If we adjust for date and latitude, the average total unadjusted time budget for the pre-300,000 populations is about 142.2 per cent at low latitudes and 139.4 per cent at high latitudes, with the equivalent figures after the 300,000 watershed being 146.7 per cent and 142.4. per cent. Given that we have managed to find solutions to the *ergaster/erectus* time overspend, it is only the difference between the latter's original time budget of 129.5 per cent and those for the *heidelbergs* that we now need to worry about. This suggests that early *heidelbergs* only needed to account for an extra 11 percentage points, while the later *heidelbergs* had to find additional savings of 13 percentage points at low latitudes and 12.5 percentage points at high latitudes.

The obvious answer is, of course, cooking. A fully cooked meat-plus-tuber diet with a 50 per cent increase in nutrient uptake would have reduced the late *heidelbergs'* feeding time requirement alone by 14.7 percentage points[6] (Fig. 6.3). That would have been more than enough to wipe out the deficit for the low latitude *heidelbergs*, while the high latitude ones would have been a comfortable 5 percentage points in credit. Since they have room to spare, it means they wouldn't have had to cook their whole meat-and-tuber diet, offering a transitional phase from casual cooking (as in *erectus*) to somewhat more regular cooking in early *heidelbergs* to a lot of cooking in the late *heidelbergs*, making the whole argument a great deal more plausible by allowing the earlier *heidelbergs* time to learn how to control fire and manage cooking. The spare capacity that would have been available to the early *heidelbergs* would have been critical in allowing the further increase in brain size that kicked off 300,000 years ago. The triangles in Fig. 6.3 give a sense of the scale of the impact that a heavy meat diet, serviced by cooking to allow a full expensive tissue effect, might have had on total time budgets.

Cooking has another unexpected benefit. It turns out, from recent studies of the control mechanisms of feeding behaviour in rodents, that the process of eating triggers the endorphin system. This probably explains why we feel contented and relaxed after eating a particularly large meal. If eating big meals, as for example at feasts, triggers endorphins, then doing this collectively, as cooking more or less

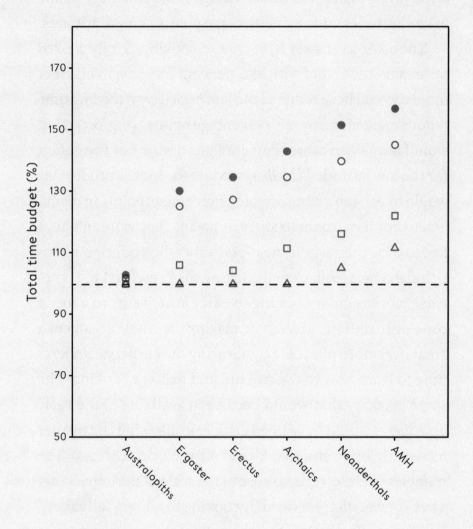

Figure 6.3

The impact of a significant meat diet on the total time budget
of the main hominin species. As in Fig. 5.6, the graph shows
the baseline total time budget (●), the reduction due to the
expensive tissue effect (○), laughter (□), and adds the impact of
a heavy meat diet with cooking (△).

forces us to do, may well have beneficial advantages in terms of social cohesion. We feel warm and friendly towards those with whom we eat. This might explain why we find social feeding so important – why we so often feel that meeting over food is a natural way of getting to know people, especially those we don't yet know well but feel we need to know better. Social eating of this kind seems to be universally important across all cultures, yet no one has ever stopped to ask why we do this. Social anthropologists speculate on the social meaning of menus, and Neolithic archaeologists get excited about evidence for feasts (see Chapter 9), but no one ever seems to ask why we go to all this trouble. They just seem to take it for granted that it is something humans do. The obvious answer is social bonding. If so, then it may have allowed archaic humans to shave some time off their social time budget, and that might have made enough of a difference to allow the increase in brain volume.

So it seems that the invention of cooking might have provided the crucial innovation that allowed *heidelbergs* to solve their particular time-budgeting crisis. In addition, it would have allowed some of the additional social time for bonding their larger groups to have been taken care of by the communal eating necessitated by cooking – although it is unlikely that this was sufficient to explain the whole of the very significant increase in community size that occurred in the *heidelbergs*. Something else was probably required, and it was unlikely to be more laughter because, given that even now we cannot involve more than three people in a laughter event, it seems already to have been at the limits of its bonding capacities in the final stages of the *ergaster/erectus* phase. I will return to this crucial point later in the chapter. First, however, I want to consider how time budgets might have worked out for the other archaic human species, the Neanderthals.

The enigmatic Neanderthals

The Neanderthals have captured the popular imagination like no other species in the story of human evolution. In the folk view, the Neanderthals were shambling, dull-witted cavemen who symbolized all that was primitive about our past: the modern dinosaur, the evolutionary failure over whom sharp-witted, inventive modern humans deservedly triumphed. In reality, the Neanderthals were a very far cry from being an evolutionary backwater – or, for that matter, dull-witted. For well over 250,000 years – longer than modern humans have been in existence – Neanderthals were successful occupants of Europe from the Atlantic as far east as Uzbekistan and Iran, and from southern Britain in the north to the Levant in the south. They coped with Ice Age conditions, and became some of the most skilled hunters of big game ever seen.

Neanderthals certainly had a very different lifestyle to modern humans. Direct evidence of dietary specialization provided by analyses of the nitrogen and carbon isotopes[7] in collagen from the bones of European Neanderthals indicate that they had nitrogen isotope levels that were significantly higher than those of the herbivores present at the same sites and of about the same magnitude as in the bones of contemporary carnivores (arctic fox and wolf), while their carbon isotope levels indicate that they were preying mainly on terrestrial mammals, and not species that lived in watery environments like fish or waterbirds (Fig. 6.4). Evidence from sites in the Levant suggests that Neanderthals made particular use of heavy spears, often tipped with triangular Levallois stone 'points'.[8] These would most likely have been used in confrontational 'ambush hunting' in which the hunters spear the prey from close quarters. Unlike the spears used later by modern humans, Neanderthal spears were not designed to be thrown like javelins, but were used more like pikes for thrusting. Their shorter arms compared to modern humans would have provided a lower leverage arc when throwing spears, so that the speed and distance over which they could have thrown a javelin-like spear would have been lower than that for modern humans, making this way of hunting less effective. Instead, the heavy build and considerable upper body strength of the Neanderthals may have given them a distinct advantage in confrontational-style hunting.

The Neanderthals exploited different prey in different locations, suggesting that they adapted their hunting strategies according to circumstances. At the Neanderthal rock-shelter of Krapina in Croatia (dated to the beginning of the last interglacial, around 130,000 years ago), most of the bones of rhinoceros and other prey typically come from juveniles, probably brought back as intact carcasses. On the other hand, those from later living floors dated to around 120,000 years ago consist mainly of skulls and horns, with very few skeletal bones, most of them from old individuals. Elsewhere, Neanderthals concentrated on adult animals: at one of the most northerly sites ever occupied by Neanderthals

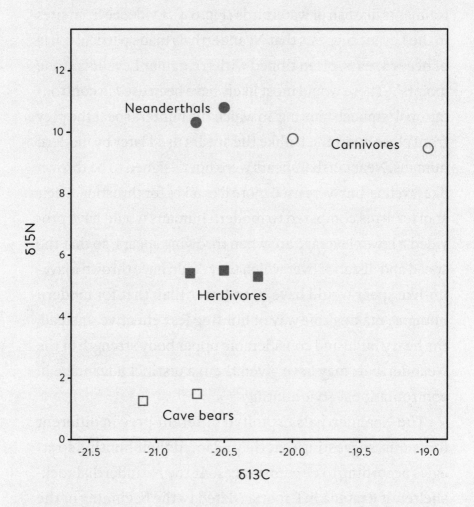

Figure 6.4

Carbon (δ13C) and nitrogen (δ15N) isotope values in the
bones of Neanderthals (●), carnivores (arctic fox and wolf: ○),
herbivores (bison and deer: ■), and cave bears (□) at the Vindija
cave, Croatia (dated to approx. 28,500 years ago).

(the German site of Salzgitter Lebenstedt), adult reindeer were hunted almost to the exclusion of everything else. Later, during the last glacial some 40,000 years ago, Neanderthals seem to have been systematically hunting medium-sized herbivores like red and fallow deer. At Bolomor in Spain, there is evidence for the regular cooking of rabbits. All this suggests that they were versatile hunters who adjusted their hunting strategy to suit the prey that happened to be available at a given location.

The fact that many of the prey animals at some sites were quite old has been interpreted as suggesting that Neanderthals were mainly scavenging the carcasses of animals that had either died naturally or been killed by carnivores (who also tend to target old or incapacitated animals). However, the absence of any tell-tale carnivore tooth marks on the bones argues against this; it is more likely that they were killing the prey themselves. Many of these animals were so large (e.g. mammoths) that several hunters would have been required for a successful kill, acting in concert to spear the luckless prey at very close quarters. It was clearly a dangerous form of hunting because, as we noted above, Neanderthal skeletons often exhibit evidence of shock damage, the conventional explanation for which has been that they were injuries received from large-bodied prey made rather angry by being speared.

The level of cooperation implied by the Neanderthals' style of ambush hunting has been interpreted as suggesting that they were fully modern in their cognitive abilities and outlook because they had to be able to coordinate with each other in order to hunt in this way. And this has, in turn, been taken as evidence that they shared with us the kinds of prosocial and altruistic behaviours that form so central a part of modern human behaviour. Indeed, the old cripple from the Neanderthal site at Shanidar in modern Iran and the old man at the French site of La Chapelle both had sufficiently serious disabilities that they would have found it difficult to feed themselves by hunting, and this kind of secondary evidence has been rather enthusiastically interpreted as implying that the Neanderthals looked after their sick and elderly group members rather than simply abandoning them.

There is, of course, no question about the fact that, with brains that were among the biggest of all the hominins, ancient as well as modern, Neanderthals cannot have been intellectual slouches. That said, their low foreheads and the distinctive bulge at the back of the head (the defining Neanderthal 'bun') have always signalled that something must be different. Analyses of the faint imprints of the brain that are left on the inside of skulls have suggested that their brains might not have been organized along quite the same lines as ours. For sure, they showed the same patterns of asymmetry (often assumed to be a marker for language) and gyrification (the folding of the brain surface) as ours, but at the same time their brain case was longer and less globular than ours, with smaller temporal lobes and olfactory bulbs, and less space for the frontal lobes. To what might they have owed these differences? And what were the implications for their social and cognitive life?

It's all in the eyes

Brain size is not everything: as we know from the social brain relationship, it is certain parts of the brain (the neocortex, and especially the frontal lobes) that are crucial in determining the size of a species' social world. Besides this, the brain consists of a number of different units, some of which do very specific tasks. Visual processing, for example, is handled in the occipital lobe at the back of the brain, and although visual input is passed progressively through a series of brain regions for increasingly detailed analysis right up into the frontal lobes (where, in effect, we attach meanings to these visual images), nonetheless a large chunk of the back of the brain does very little else except process the incoming stream of messages from the retina.

And this perhaps gives us a clue about what Neanderthals might have been up to: is the 'bun' at the back of their head that so defines them telling us that they had an unusually well-developed visual system. But if so, why? It turns out that they were indeed more visually oriented than we are, but for reasons that are a bit 'left field'. Let me explain.

The Neanderthals faced a novel problem that was unfamiliar to any of their tropical predecessors: short winter days and low light levels, even in summer. Within the tropics, daylength hardly varies at all and, barring the odd cloudy day, sunlight is bright and all-illuminating. But the further north and south you go out of the tropics, the more seasonal daylength becomes (thanks to the tilt of the earth relative to the sun, winter days are short while summer days are long) and the weaker the sunlight is (because it is filtered through increasing layers of the atmosphere, and thus loses much of its strength, which of course is why it is also colder at high latitudes). Winter is likely to have been a particularly challenging time for them, not just because of the cold but because all their foraging (and presumably social activity) had to be crammed into a very short day: mid-December daylength is as short as 10 hours at 30° latitude (the north coast of the Mediterranean) and just 8 hours at 45° latitude (about as far north as the Neanderthals ever got even during the interglacials). At the higher latitudes, they were losing as much as 4 hours each day (one third of their active day!), yet still having to balance their time budgets in terms of actual numbers of hours devoted to each activity. Feral goats (who are strictly diurnal, at least in northern Europe[9]) face exactly this same problem: our model of goat time budgets and biogeography shows that their ability to occupy high latitude habitats is dramatically affected by shortening winter daylengths as they go further north in Europe.

The Neanderthals would also have faced another problem when it was daylight: the light quality is much poorer at high latitudes and this would have meant that they couldn't see things in the distance so well. For a hunter, this is a serious problem, because you really don't want to make the mistake of not noticing the mother rhinoceros skulking in a dark corner of the forest edge when trying to spear her calf. Living under low light regimes places a much heavier premium on vision than most researchers imagine.

The evolutionary response to low light levels is to increase the size of the visual processing system. It is the familiar principle from conventional star-gazing telescopes: under the dim lighting of the night sky, a larger mirror allows you to gather more of the light from whatever you want to look at. By the same token, a larger retina allows you to receive more light to compensate for poor light levels. A bigger retina inevitably necessitates a bigger eyeball in which to house it – nocturnal primates, for example, have much larger eyeballs than diurnal ones. And then there is no point in having a super-large light-receiving mechanism (the eye) if you don't have a big enough computer behind it to process and integrate the extra information. Because the visual system is hierarchically organized, its successive layers are closely scaled with each other – from the retina through the optic nerve, the optic chiasma and lateral geniculate nucleus way stations, to the primary visual cortex (V1) at the very back of the occipital lobe, and so on up through the layers of the visual system (V2 through V5) into the temporal and parietal lobes.

So here is the smoking gun: Neanderthals had eye sockets that were around 20 per cent larger than those of anatomically modern humans living in the same habitats. Might it be that they developed an unusually large visual system to cope with living under low light regimes at high latitudes? And if so, was their large brain disproportionately devoted to vision, and hence less well endowed with those regions at the front of the brain that are so important for social cognition?

The answer appears to be yes, but this conclusion comes from putting several different bits of the jigsaw together, so the answer isn't straightforward. One of these pieces is that, surprisingly, this relationship holds even in modern humans. Ellie Pearce, then one of my students, measured the skulls of museum specimens of historical modern humans from different parts of the world. She showed that eye socket volume correlated with cranial volume across these populations, and that both correlated with latitude: populations that lived at high latitudes had larger brains and larger eye sockets than ones that lived near the equator. In a follow-up study, she showed, using brain scans from people from different parts of the world, that this was also true of the size of the visual areas in the occipital lobe itself. Yet visual acuity (in effect, the ability to read small print) under natural light conditions (as opposed to the optician's artificially lit clinic) remains constant across latitudes for populations born and reared at those latitudes. So it seems that populations living in different latitudes have the same visual acuity, but require ever larger visual systems to achieve this as they move pole-wards. In other words, even modern humans seem to compensate for the reduced light levels at high latitudes by increasing the size of the visual system in order to maintain visual acuity at a roughly constant level – in this case, without sacrificing those parts at the front of the brain that determine their intelligence and social skills.

The key here is the correlation between orbit size and the size of the visual system. If we can use this relationship to estimate visual brain volume for fossil species, we can remove this part of the brain and get a better estimate of the size of the brain region that actually determines community size. As we saw in Chapter 3, the primate data suggest that the best predictor of social group size is in fact the frontal lobes of the brain, but we don't have any principled way of estimating this from fossil crania. However, if we can remove the visual system we at least remove a major chunk of the brain that doesn't correlate particularly well with group size, and that should sharpen up our estimates considerably.

If we apply the relationship between orbit size and visual system size to Neanderthals and other archaic and anatomically modern fossils in Europe to estimate the principal regions of visual cortex, and then subtract this from total brain volume we get a better estimate of the social brain for each species. Interpolating these residual brain volumes for individual fossil specimens back into the ape social brain equation gives us Fig. 6.5. Neanderthals would have had communities that were significantly smaller than those of contemporary AMH (and a great deal smaller than we might have imagined had we not corrected their brain volume for the size of their visual system). In fact, as the figure shows, Neanderthal community sizes were identical to those of *heidelbergs* (about 110 individuals in size,[10] so about two-thirds the size of AMH communities). In other words, as Neanderthals became adapted to high latitudes, they increased visual system size but not frontal lobe size. Anatomically modern humans started off with community sizes that were virtually identical to those of *heidelbergs* (their parent population), but underwent a progressive increase in group size (and correlated brain evolution) during the early phase of their evolution in tropical Africa where they had yet to experience the problem of low light levels.

The difference in group size between Neanderthals and AMH is highly significant, and has major consequences for Neanderthal social time budgets. This does not affect their

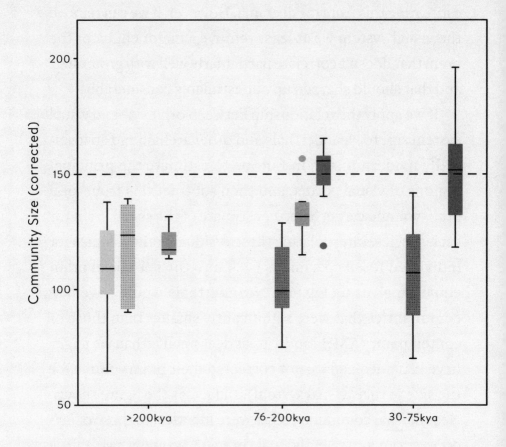

Figure 6.5

Median (with 50 and 95 per cent range) community sizes for Middle and Late Palaeolithic hominins, obtained by interpolating individual cranial volumes into the ape regression equation from Fig. 3.1. The cranial volumes for archaic and Neanderthal specimens have been corrected for the effect of latitude on the visual system (following Pearce et al. 2013). Note that *Homo erectus* cranial volumes have not been corrected for latitude.

    Heidelbergs

    Archaic Neanderthals

    Neanderthals

    Archaic Modern Humans

    Anatomically Modern Humans.

feeding time costs, of course, because these are determined by overall brain and body size. However, a difference of around 40 in social community size compared to modern humans would have saved the Neanderthals 11 percentage points on their social time budgets, and this would have reduced the pressure on them enormously. They would still have had to devote 67.5 per cent of their day to feeding (Fig. 5.2), but the net effect is to bring their overrun down from 39 per cent (adjusted for the expensive tissue effect) to 28 per cent, and so in line with the budget overrun for the other *heidelbergs* (Fig. 5.3b). And that means the same solutions as the *heidelbergs* exploited to solve their time-budgeting crisis would have satisfied the Neanderthals' requirements. There was probably enough leeway in the time savings from cooking to have allowed the little extra time needed for the Neanderthals to increase brain size by the modest amount they did over their *heidelberg* ancestors (141 cc, or 12 per cent).

The singing Neanderthals?

The British archaeologist Steven Mithen has famously written at considerable length about the 'singing Neanderthals', arguing that they (and modern humans) evolved vocal and speech capacities that were essential to their survival during the challenges of the Ice Age. Even though Neanderthal (and *heidelberg*) communities must have been smaller than those of anatomically modern humans, they were nonetheless sufficiently large to raise doubts about the adequacy of grooming and laughter as effective bonding mechanisms. They must have had another strategy to help fill the gap.

There is one possible solution, namely music. Singing has an obvious continuity with laughter. It involves exactly the same anatomical and physiological processes: they are both all about breath control, are hard work for the chest wall muscles and diaphragm, and are consequently likely to be an effective mechanism for endorphin activation. Wordless singing (or humming) also shares with both laughter and speech a number of properties, including segmentation, articulation, phrasing and synchrony, and this makes it an ideal transition between laughter and full speech. I will return to music in more detail in Chapter 8. For the moment, I just want to highlight the physical activities we associate with music: singing (without words), dancing and rhythmic music-making (drumming, clapping, playing musical instruments of various simple kinds).

There is a possible precursor for this in primates. Gelada baboons live in modest-sized reproductive units (harems that consist of a breeding male, sometimes an additional follower male, and 3–6 reproductive females and their offspring) that associate together in bands which average around 100–120 individuals (very much larger than the groups of any other primate by a considerable margin). The presence of so many other animals in these large groupings must impose a significant stress on social bonding within the harem units, and gelada devote a very considerable amount of time (the most of any wild primate so far recorded) to social grooming. However, in terms of the grooming time relationship (Fig. 2.1), they do not spend anything like what they ought to for their natural group size. With an average band size of about 110, they ought to spend 36 per cent of their day grooming according to our grooming time equation, but in fact they spend only about 17 per cent on average. Their solution to this dilemma seems to have been to develop a form of vocal grooming.

Gelada are among the most vocal of all the primates, with a repertoire of sounds that is both larger and more complex than that of any other monkey (or ape). In particular, they have an unusually complex set of contact calls (especially so compared to other baboons and macaques) and use these calls constantly not only during grooming interactions but also, more importantly, when they are feeding. The calls are used both in antiphonal call-and-reply sequences between grooming partners (rather in the manner of human conversations) and in choruses when all the adults in a harem call together. The timing of these calling bouts is very fine-tuned, with individuals calling in sequence with intervals that are so short they cannot just be responding to each other's calls but must be anticipating other individuals' call patterns and timing their own calls to slot in appropriately. These calls seem to be functioning as a form of vocal grooming-at-a-distance that, in effect, allows grooming partners and harem members to continue interacting with each other when the business of the day (feeding or travel) makes physical contact impossible.

What makes these calls particularly interesting in the present context is that they have decidedly musical properties. They have a sing-song quality that is unique among the primates – as perhaps befits the fact that gelada are also unique in habitually living in unusually large social groups by primate standards. Producing these varied calls requires that gelada integrate control over the vocal space (lips and tongue, and the mouth cavity) in a way that seems not only to be unique among the non-human primates but also rather similar to that used by humans to produce speech. As such, it provides a perfect model for how chorusing might have got going in the hominin lineage.

If music is to act as an extension to grooming and laughter, then it must share with these the capacity to trigger endorphins. We tested this by running a series of experiments that involved various kinds of musical activity using the same design as we did with laughter (see Chapter 5). We used change in pain threshold as our test for endorphin activation and compared groups who did a musical activity (a religious service with singing, a drumming circle, dancing) with control groups who did not. The results suggested that doing music in any form generated an endorphin surge, whereas a more static activity or passively listening to music did not. So it looks like music can also be used to trigger the pharmacological mechanism that underpins social bonding.

Music-making has a couple of important advantages over laughter as a mechanism for social bonding. One is that it can involve many more individuals, thereby radically enlarging the 'grooming' circle. We do not, as yet, have any idea what the upper limit on the effectiveness of musical performance is, but it will certainly be greater than the three individuals that seems to be the upper limit for laughter. But so long as its effective group size is larger than three, it will enable a proportionately larger group to become bonded. The other advantage is that music-making (whether singing, playing instruments or dancing) is an activity that involves close synchrony, mainly of course engineered by the beat that keeps everyone in time.

There is something genuinely odd about synchrony, because it seems to ramp up the endorphin production generated by physical exercise by something close to a factor of two. We demonstrated this in a rather elegant study carried out on the Oxford University boat crews by Robin Ejsmond-Frey (then president of the Boat Club and already a triple Blue) and Emma Cohen. Rowing at this competitive level has little to do with physical strength: the crews are all about equally strong and equally fit as each other. What really makes the difference between winning and losing is how long they can keep up their strike rate and their timing. If a crew of eight oarsmen starts to get out of time with each other, they lose a great deal of the thrust that propels the boat through the water. This makes sweep-oar rowing an ideal test-bed for looking at behavioural synchrony.

We tested for endorphin production in crew members using the change in pain threshold, and we did this while they were doing their early morning training on rowing machines in the gym. First we had them work out on the machines alone and then, on separate days, they did it with the machines hooked up together as a virtual boat. When hooked up as a virtual boat, the crew members have to row in synchrony, as they would do on the water. The results were spectacular: endorphin activation in a virtual boat (as indexed by the change in pain threshold) was double that when the oarsmen were rowing alone, even though the power output in both conditions was identical. Somehow, behaving in close synchrony seems to ramp up the endorphin effect by a very considerable margin.

If the role of music (perhaps initially in the form of laughter-like wordless chorusing, but later including more familiar forms of music-making and dancing) was to engineer a form of grooming-at-a-distance that could encompass more layers of the social network once human community sizes began to edge up above the 75 individuals typical of *ergaster* and *erectus*, then the added social time requirement might have been quite modest compared to what would have been needed if each person had to be contacted individually.

We can get a sense of the scale of the problem by looking at how we bond our own social networks. In contemporary human networks, the time investment per individual declines rapidly as one passes out through the layers of the social network (Fig. 3.5). We only see each of the 100 individuals in the outer 150 layer of our social networks approximately twice a year, whereas, on average, we see each of the members of our innermost circle of 5 every other day. Between them, these five people account for around 40 per cent of all our social effort – and our emotional capital. The outermost layer accounts for less than 20 per cent of our social effort, but that still means 200 individual contacts a year and, even if each was only short, between them they would occupy a significant amount of time – especially if they involved travelling to see the individuals concerned, as is likely to be the case for people in this layer. In both hunter-gatherer and modern societies, most members of this outer layer live at least a day's journey away: the mean distance to members of the outer circle in the sample of European personal social networks in Fig. 3.5 was 17.8 *hours* – the better part of two days' travel away.

Of course, people live in families, so most such visits might allow you to tick off an average of, say, five people at a time. Nonetheless, even allowing a single day per visit, the 100 people in the outer 150 layer would still require you to spend 38 days a year just visiting your more distant friends and relatives – never mind having to see the 50 people in the more important inner layers, some of whom require near-daily contact. If musical performance allowed you to do this all at once in a single annual dance, it would significantly reduce the time cost since only one or two days need be devoted to 'grooming' with all of them. Even if one allowed for a three-day feast with singing and dancing, it would still save 35 days if such an event was held just once a year, effectively giving you the whole of the outer layer almost for free. The 6 per cent extra social time required by *heidelbergs* to service their larger community size would translate into 22 full days across the year; if this was reduced to three days, then the saving is clearly substantial and would reduce the required extra allocation for social interaction to a mere 1 per cent above the *ergaster* baseline. The *heildelbergs* would be well within their time budgets.

The only fly in the ointment here, however, is that arranging annual events of this kind surely requires language. And this raises the question of whether any of these archaic humans had language – or, at least, language of sufficient complexity to allow a community to negotiate such events. I want to postpone discussion of just when language evolved to the next chapter. But in the meantime, it is important to place a marker of doubt against this mechanism until we have done so. Nonetheless, music might well still have evolved in archaic humans, perhaps as a supplement to laughter in bonding more modest-sized groups, even if not as a community-wide bonding mechanism.

We have managed to solve the time-budgeting crises faced by archaic humans mainly by finally introducing cooking, but also partly by introducing music-making. The archaeological evidence demonstrates that there was a key watershed at around 300,000 years ago that released a major constraint on brain size, just about the time the Neanderthals got going in Europe, and I have suggested that this coincided with the appearance of cooking and the control of fire. We must now ask the same question of anatomically modern humans – our fourth transition.

The Fourth Transition

MODERN HUMANS

Around 200,000 years ago, a more gracile form of hominin emerged in Africa – anatomically modern humans (AMH) or *Homo sapiens* had finally arrived. The somewhat fragmentary fossil evidence suggests that this new form replaced the archaic humans in Africa quite rapidly: by 100,000 years ago, there were no archaic populations left, although what had become of them remains unclear. Elsewhere in mainland Europe and Asia, however, archaic humans remained the only hominins, represented variously by the Neanderthals, the Denisovans and some remnant populations of *Homo erectus* in the Far East. By 70,000 years ago at the latest, this new human species began to move across the Levant land bridge that connects Africa with Eurasia, with successive waves of emigrants heading east across both the top and the bottom of the Red Sea at different times (Fig. 7.1). This exodus from Africa at 70,000 years ago is now known as the 'Out of Africa' event.

Those who tried the northern route through the Levant would certainly have bumped into the Neanderthals. Indeed, it may well be that the presence of Neanderthals in the Levant initially prevented modern humans using this northern route out of Africa. The earliest exoduses were probably

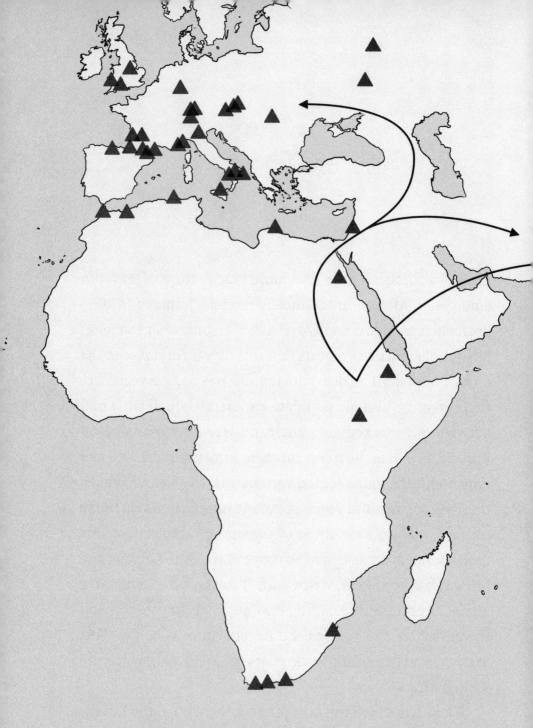

Figure 7.1

Distribution of anatomically modern human fossil sites. The arrows denote the early route out of Africa across the Strait of Hormuz into southern Asia and the later route through the Levant into central Asia and, eventually, Europe.

After Klein (2000), Bailey and Geary (2009) and Osaka City University (2011).

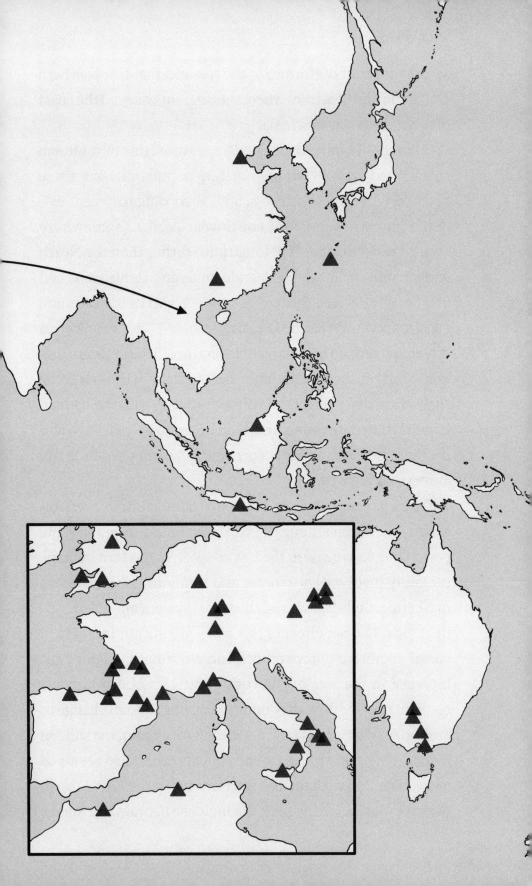

across the Strait of Hormuz at the Neanderthal-free southern end of the Red Sea, from where the migrants hugged the coast of Arabia into southern Asia.

Like the Neanderthals and Denisovans, this new human species had evolved out of *heidelberg* populations, but these had been African *heidelbergs*, and, according to recent evidence, most likely from southwest Africa (somewhere around 14°S latitude, 12°E longitude) rather than the North African ones from which Neanderthals and Denisovans had presumably evolved (hence the depth of AMH's last common genetic ancestor with the Eurasian archaics). This new species was characterized by a marked skeletal gracilization and a dramatic further increase in brain size – although there is some suggestion that this was a two-part process, with the skeletal gracilization happening first around 200,000 years ago and the increase in brain size occurring later, perhaps even as late as 100,000 years ago.

The increase in human brain size paralleled (or even followed) that seen among the Neanderthals during the same period but, as we saw in the last chapter, in modern humans it mainly involved the frontal and temporal lobes whereas in Neanderthals it involved the visual system and the occipital lobes. The net effect of this was a significant increase in social cognitive competences and a dramatic 36 per cent increase in the size of the communities modern humans could maintain. Why they underwent this sudden change in brain size is, however, far from clear. Once again, the sudden increase in brain size at around 100,000 years ago seems to have coincided with further fluctuations in the East African rift valley lakes. In this case, the increase in brain size seems

to have coincided with the drying of these lakes, perhaps forcing early human populations up the corridors created by the
lakes in a search for better-quality habitats. Whether larger
brains were necessary to allow for better foraging skills or
with the need to have large extended networks remains an
open question.

Two important consequences followed from the appearance of AMH. One was that the species rapidly spread
throughout the habitable planet: by 40,000 years ago (barely
30,000 years after first leaving Africa), modern humans had
colonized Australia, and by 16,000 years ago at latest they
had romped their way down the whole of the American
super continent from their entry point in Alaska at the very
top to the southernmost parts of the southern continent.
The speed at which modern humans occupied the Americas
is little short of miraculous, given that it was done on foot
long before the domestication of any transport animals.[1] The
second consequence is that the arrival of modern humans
in Europe some 40,000 years ago seems to have coincided
with the subsequent disappearance of the Neanderthals. The
Neanderthals had successfully occupied Europe and western
Asia for the better part of 200,000 years by the time modern
humans turned up – like all subsequent historical invaders –
from the Russian steppes to the east. Within 10,000 years
of the arrival of modern humans, the Neanderthals had van-
ished. Why they disappeared has remained one of those
iconic mysteries that have continued to fascinate us – not
least because they died out less than a thousand generations
ago, so recently that it seems like only yesterday. Were they
murdered by the invaders from Africa, or was it just bad luck

that they were less able to cope with the exigencies of the last great Ice Age?

I will return to this question at the end of the chapter. First, we need to examine this new species, modern humans, in more detail and then look at why they were so successful.

The history in the genes

Molecular genetics has opened up wonderful new vistas in the study of population history in the past two decades. By comparing the detailed molecular structure of DNA in living people, it has been possible to reconstruct in considerable detail the history of modern human migrations. Perhaps the most surprising finding has been the fact that all modern Europeans, Asians, Pacific islanders, Native Australians and Native Americans are much more closely related to each other than they are, on average, to any of the peoples of Africa. This implies that they all descend from the same small subset of the peoples of Africa. It was this evidence that first drew our attention to the migrations out of Africa around 70,000 years ago: the small amount of genetic variation among the non-African races has a coalescence time (the point at which their various new mutations would converge on a single common ancestor) at around 70–100,000 years ago.

Most of the genetic analysis for this has been done on mitochondrial DNA (mtDNA), which is inherited *only* through the female line (men and women all have their mother's mitochondria). This means that when you look at someone's mtDNA you are seeing a direct uninterrupted lineage right back to some deep female ancestor. The Y chromosome, which is inherited only by males from their fathers, shows a very similar pattern, but has a much shallower convergence date (closer to 60,000 years ago), reflecting the consequences of polygynous mating (which unusually involves rather fewer males siring offspring than when monogamy is the norm).

There are four major mtDNA haplotypes (genetic lineages, or families) in Africa, reflecting the longer history of modern humans on that continent and a geographical dispersion that favoured genetic divergence. The four mtDNA families are known as the L0, L1, L2 and L3 haplotypes, each with many subfamilies of its own. The first three of these are found, respectively, in southern and eastern Africa (associated mainly with ancient click-speaking[2] hunter-gatherer peoples like the Khoisan and the Hadza), in western and central Africa (associated in particular with the pygmy peoples), and in western and southeastern Africa. L2 is the most common of all the haplotypes. The L3 lineage is the youngest (i.e. most recent); it is associated in particular with the Bantu peoples (who originated in East Africa but underwent a major historical expansion into west Africa and, much later, a second major expansion down into southern Africa starting around 3,000 years ago), as well as with most of the Semitic and semi-Semitic peoples of north Africa and the Arabian peninsula (including, incidentally, the Jews). The only two mtDNA haplotypes represented outside Africa (known as the M and N macro-lineages, to which all European, Asian, Australian and Amerindian peoples belong) are descended from one particular subset of the L3 haplotype that was historically present in eastern Africa.

The genetic data suggest that, while the three most ancient haplotype lineages have undergone slow steady expansion over the last 150,000 years, the L3 haplotype underwent a dramatic and sustained population explosion from around 70,000 years ago, quite soon after its first appearance as a new mutation in eastern Africa (approximately 90,000 years ago). In contrast, the only other haplotype to show evidence for a population explosion (the L2 haplotype) did not undergo its expansion until around 20,000 years ago. Whether the demographic explosion in the L3 lineage occurred before the Out of Africa event (and so was its cause as people tried to find unoccupied land in which to live) or afterwards (as a radiation due to release from ecological constraint following emigration into unoccupied land in Asia) remains to be seen – the timing of these events is insufficiently precise to be able to tell for sure which came first. Nonetheless, it is clear that the rapid expansion in the L3 lineage coincides with, and is hence intimately related to, the first migrations out of Africa by modern humans.

An ingenious solution to an impossible problem

Anatomically modern humans had the same brain size as the Neanderthals. However, unlike the Neanderthals, AMH had an additional social time component to cope with since, as we saw at the end of the last chapter, the social brain relationship suggests that they had community sizes that were a third larger than those of archaic humans. That alone would have committed them to finding an additional 12 percentage points for their social time allocation, and pushed their time budget up to an impossible 150 per cent of the day. How did they solve their time-budgeting crisis?

In Chapter 6, I suggested that the Neanderthals would have had to cook as much as half their diet to balance their time budget and we know that meat did become a major feature of their diet. One possibility, then, would have been for modern humans to have increased still further the amount of cooking (and hence meat) so as to free off enough time for the extra socializing they had to do. To find the extra 12 percentage points they needed to save on feeding time, modern humans would have had to have an additional 18 percentage points that benefited from cooking – equivalent to an additional 20 per cent[3] of their diet on top of the 42.5 per cent that the Neanderthals were already cooking. In other words, around 42.5 + 20 = 62.5 per cent of their diet would have had to consist of meat and tubers. That seems implausibly high, given that only around 45 per cent of the modern hunter-gatherer diet actually benefits from cooking (Table 5.1). The proportion of meat in the diet of modern hunter-gatherers varies rather little across latitude, and typically accounts for only 35–50 per cent of the total diet. It is only at very high latitudes (above 60°) that modern human populations become heavily dependent on meat, and much of that is fish (which is easier to digest even when raw, as anyone who has tried Japanese cuisine will know). Modern humans probably didn't reach such high latitudes until after 10,000 years ago, long after they had evolved their big brains. Since modern humans are not adapted to be pure red meat carnivores, this option seems at best unlikely.

If modern humans didn't gain their savings by cooking more of their diet, where did they come from? At the end of the last chapter, I raised the question of whether the archaics introduced communal singing and dancing as a way of increasing the circle of people that could be bonded at the same time. I suggested that they might have used this to lift the bonded circle from the 75 or so characteristic of early *Homo* to the 100 or so characteristic of the archaics. Although music and dance would certainly have been functional in bonding the local community (the band or camp group), and might well have evolved to reinforce the bonding effects of laughter, it seems unlikely that communal dancing would have worked on the larger community scale in the absence of language – or, at least, of language that was sufficiently complex to make the necessary arrangements.

With language allowing periodic communal dances, this could have reduced the additional time requirement for bonding the extra members of the very large (compared to *ergaster*) modern human communities of 150 to about 1 per cent of the day. However, remember that, in primates as a whole, grooming time is concentrated on close friends while at the same time being proportional to the total size of the group or community (see Chapter 2). This is because the need to invest in your close allies is proportional to the stresses experienced from the rest of the community. Communal dances might allow you to reduce the time you invest in peripheral members of the community, but it doesn't necessarily help you solve the increased time you need to devote to your close friendships in order to ensure that they support you and come to your aid when you need it.

Language has considerable advantages over grooming as a bonding mechanism because it allows more efficient communication. Its efficiencies include allowing us: (1) to interact with several individuals at once;[4] (2) to time share on other activities (we can talk while walking, cooking and eating, all of which are incompatible with grooming); (3) to acquire information about the state of the social network (on a scale that is impossible if such knowledge depends, as it does for monkeys and apes, on personal observation); and (4) to promote our interests (by advertising our good qualities or denigrating those of other individuals). These are all very effective uses of language and they could well enable significant savings on time. However, they could equally well all be emergent properties of language – novel ways language can be used *once you have language in place for some other reason*. More importantly, conversations, of themselves, do not – at least, so far as we know – trigger the endorphin release that seems to be so essential for bonding.

Yet there is one possibility that has been almost completely overlooked in the discussions of human evolution, namely what happens when we combine language with fire. Conventional discussions of fire in the archaeological literature consider only two functions of any importance: cooking and warmth. Fire was undoubtedly crucial to the evolution of cooking, and it had a long if somewhat intermittent history for this purpose through at least part of the *ergaster/erectus* phase and a more focused and certain history through the *heidelbergs*, especially after 400,000 years ago. Fire must also have played a seminal role in alleviating the hardship of living at high latitudes where, especially during winter night temperatures are significantly lower than they are in the tropics. The increased control over fire after 400,000 years ago must surely have made it possible for archaic and modern humans to survive in Europe and western Asia as the ice sheets advanced and dramatically lowered temperatures.

However, fire has another big advantage: it is an artificial source of light that can be used to extend the active day. This would obviously be an important asset at high latitudes during winter, when the length of the day can be as short as 8 hours. But even in the tropics, extending the active day beyond the conventional 12 hours would have significant benefits by reducing the pressure on time budgets. For monkeys and apes, the night is dead time: all primates have poor night vision (by comparison with most other mammals) and so, once it gets dark, the best thing they can do is sleep. That at least allows them to hide from nocturnal predators. But if fire provides extra time that can be used for more sedentary activities like making tools or, perhaps more importantly, socializing, then it would free up the day for more mobile activities like foraging.

Modern humans spend about eight hours in sleep, and if we take this as the minimum required, then it allows another four hours to be devoted to other activities around the campfire. Adding four hours to the length of a 12-hour active day in effect allows hominins to run a time budget that is 133 per cent rather than 100 per cent. This would absorb a major component of the excess time budget predicted for modern humans, without their even needing to cook as much of their food as the Neanderthals. Fig. 5.4 suggested that modern humans would have an overshoot of around 45 per cent on their time budget (adjusted for the expensive tissue effect). Increasing the active day by a third would leave just 145 – 133 = 12 percentage points to be found elsewhere. More importantly, it would come very close to absorbing the entire social time requirement for AMH (which Fig. 5.3a estimates as 35 per cent). The effect of extending daylength on the total time budgets of the three human species is shown by the solid triangles in Fig. 7.2.

This might seem to imply that AMH could have solved their bonding problem without needing language. However, there are two issues that need to be considered. One is whether the whole of the evening could actually be devoted to social interaction; the other is whether the additional time devoted to old-fashioned one-on-one primate grooming would have been adequate to bond their much larger communities. Other activities – notably cooking and, especially, eating – also need to be done during the evening, since they are much more efficiently done communally (or at least in family groups); although we can talk while we eat, the time-sharing isn't likely to be a perfect double use of time.

More importantly, not everyone is busy interacting during active social periods. Gelada baboons, for example, have a very distinct social period first thing every morning before they start foraging, and it can last several hours. However, our data show that only about 40 per cent of the animals are actively involved in social interaction (either as groomer or groomee) at any given time during this morning social period; the rest are sitting idle, resting, dosing, occasionally

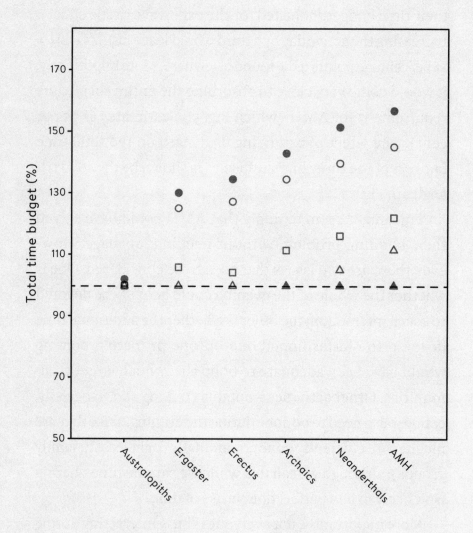

Figure 7.2

The impact of the use of fire to extend daylength on the total time budgets of the three human species. Symbols as for Fig. 6.3, with extended daylength included as filled triangles (▲) for archaics, Neanderthals and AMH only.

taking the odd mouthful of food – in effect, waiting for someone to come and groom with them. If we allow the same ratio for modern humans round their fire, then they would manage to accommodate around 14 percentage points of their social time budget in the evening, and would have to find the balance (21 percentage points) from their daytime activity budget – or by using laughter and music to compress social time. Of course, this assumes that all conversations are as intense and socially engaging as dyadic interactions that involve just two people (much as grooming does), and everyday experience suggests that they might well not be.

Nonetheless, if we assume for the time being that conversations across the fire are as intense as dyadic grooming, the crucial issue is: how big do conversations get? Our data on conversation groups consistently suggest that there is a limiting size to human conversations at four individuals – a speaker and three listeners – with an average size that is closer to three people. If language was as effective for bonding as grooming is, that would make conversation between two and three times more efficient than grooming in terms of the number of people a 'groomer' can reach. Doubling the reach of a grooming interaction would allow AMH to double its equivalent social time allocation to 28 per cent, and leave only a small deficit of around 7 percentage points to be found during the daytime. The merit of this solution is that it would mean that we wouldn't need to assume more cooking to save on the feeding time budget than is already the case for archaic humans. This is especially important, given that we needed to assume the maximum amount of cooking (by modern hunter-gatherer standards) to solve the *heidelbergs'* and Neanderthals' time-budgeting problems.

The big question here, then, is whether conversation is as effective for social bonding as grooming. In other words, where does the endorphin surge come from during conversations? I suggested in Chapter 5 that the *ergaster/erectus* lineage evolved laughter as a mechanism for using social time more efficiently; and in Chapter 6 I argued that archaic humans evolved singing and dancing as a way of widening the bonding circle still further. Presumably, all of these could be relegated to the evening hours – and, indeed, in traditional societies dancing is almost universally an evening activity. Even for us, dances don't have quite the same magic in daylight as they do at night. In fact, it seems that there is something psychologically rather special about the evening even for story-telling.

However, dancing is not something one does every day, and it may be that its slightly less intimate nature means that it is only good for the weaker relationships in the outer layers of the network. Given that primates show a distinct pattern of reinforcing their inner core relationships when group size increases (Chapter 2), maybe we do need something else that would allow these more intimate relationships to be given more time each day. Sitting around the campfire in the evening engaged in more intimate social activities like old-fashioned grooming, gossiping or laughter might still be important.

Sitting around a campfire in the semi-dark adds a significant constraint on what can be done visually, so a vocal channel has obvious advantages if you are going to make the most of the opportunity it offers. Laughter is all very well, but everyday experience suggests that one doesn't get quite the same buzz when observing laughter at a distance – hearing people laugh across the room generally elicits puzzled interest but almost never uproarious laughter. If a vocal channel is to play a significant role, then it has to involve a great deal more than simply observing someone else's behaviour. There would thus seem to be a significant selection pressure for language at this point: language would have allowed considerable information exchange in a context where a visual channel does not function especially well.

In this respect, there are two particular aspects of language that have potential advantages in social bonding. One is that language can be used to manipulate laughter by telling jokes: without language, laughter is a form of chorusing that is essentially a response to particular events, much as slapstick comedy still triggers laughter for us. However, such events occur at unpredictable intervals and are entirely dependent on vision, and hence most likely to be something that happens during daylight. Language allows us to manipulate both the frequency and context of laughter through the use of jokes, bringing it under our control and allowing us to radically increase its frequency and effect. The second important feature of fireside conversation is that it offers the opportunity for story-telling. Stories are important for two reasons that bear directly on the business of bonding the extended community. One is that they enable us to construct social histories, and so emphasize how we came to form a community united by a common history; the other is that we can tell stories about the unseen world – the world of fiction and the world of spirits, thereby making both fiction and religion possible.

I shall have a great deal more to say about both of these in the next chapter. For the moment, I just want to focus on *when* language evolved, not least because the range of the times that have been suggested is almost as wide as human evolution itself. At one extreme, the neurobiologist Terry Deacon suggested that it coincided with the appearance of *Homo ergaster* around 1.8 million years ago (his symbolic revolution), while at the other end some archaeologists have suggested that it was as recent as 50,000 years ago (coinciding with what became known as the Upper Palaeolithic Revolution in Europe with its dramatic explosion of miniaturized tools and craftwork, its cave art and statues). Notice that both of these (and most other archaeological views) focus on the symbolic use of language (being able to create verbal symbols that stand for important everyday concepts). In my view, these more sophisticated uses of language arose much later (quite conceivably as late as 50,000 years ago, as the archaeologists suggest) and are parasitic on more everyday social uses of language for social exchanges and story-telling.

If the evidence suggests that language is very ancient (e.g. dates back to early *Homo*), then my explanation here won't hold water. Similarly, if it is very late (e.g. dates back only 50,000 years), then it won't have been of any use in solving AMH's bonding crisis either. For the hypothesis to be given any support, language needs to evolve with or just before the first appearance of anatomically modern humans.

When *did* language evolve?

Determining when language might have evolved has always been rather difficult. Archaeologists have focused either on evidence for symbolism (on the reasonable grounds that it doesn't make sense to have symbolic constructs like *gods* or *ancestors* if you can't talk about them) or on evidence for brain lateralization. In modern humans, language function is localized on the left side of the brain, which also happens to be larger than the right – though whether that has *anything* to do with language function is in fact quite another matter. Be that as it may, palaeoanthropologists have assumed that evidence for brain lateralization (left side bigger than the right) is evidence for language, and have searched somewhat desperately for signs of brain lateralization on the insides of fossil crania.

In fact, both the symbolism and the lateralization claims are problematic. Convincing archaeological evidence for symbolism (mainly cave paintings and figurines, interpreted as symbolically representing or 'standing for' some real life person or concept) only appears in the Upper Palaeolithic (mostly after 30,000 years ago). All it can really do is give us a latest possible date: people might have been chatting away symbolically long before anyone thought of turning symbols into physical objects. On the other hand, lateralization may have as much to do with control over the throwing arm as anything else.[5] In fact, lateralization may be much more ancient and widespread, having evolved very early in vertebrates (it is present in prehistoric sharks) and merely become exaggerated in hominins. Perhaps genetics can help us out? If we can identify the genes that are involved in language, then we might be able use the sophisticated statistics of evolutionary genetics to estimate when these particular mutations first arose. The FoxP2 gene (mutant alleles associated with speech and grammar deficits in living humans) and the myocin gene MYH16 (associated with large jaw muscles in the apes, but inactivated in modern humans with our small ones) have attracted particular attention in this respect.

Initial estimates suggested a date of origin for FoxP2 of around 60,000 years ago – which obviously made some of the archaeologists very happy since it very conveniently just predates the first evidence for symbolic art in the Upper Palaeolithic. However, FoxP2 was later found in the Neanderthal genome and interpreted as evidence for language in this species. That would suggest an origin for FoxP2 at around 800,000 years ago at the point when the Neanderthal and AMH genetic lineages parted company – implying, of course, that all archaic and modern humans (and perhaps even some late *erectus* populations) had language. However, a similar gene was subsequently found in birds, suggesting that its real function may have more to do with control over vocal articulation rather than language as such. In other words, it may be that FoxP2 tells us about speech (or, plausibly, singing) and not language.

Estimates for the time of origin of the human form of the myocin gene give a value of 2.4 million years ago, leading to the claim that language evolved in very early *Homo* or perhaps even their late australopithecine precursors. The real question, however, is what the myocin gene is actually telling us. In reality, it is difficult to see the myocin gene as having anything to do with language at all: smaller jaw muscles might conceivably be necessary for language (and laughter, for that matter), but they are certainly not sufficient. They are much more likely to be associated with a change of diet – which we can at least vouch for at precisely that point in time in terms of the switch to increasing use of meat in early *Homo* (Chapter 5). More likely, if small jaw muscles are necessary for language (and that has yet to be demonstrated), then it is more likely that they are just one of the many things that had to be in place to make it possible for language to evolve at some much later date.[6] They are not markers for language, but essential precursors for language.

There are, however, three other approaches we can use to try to pinpoint the origins of language. One is neuroanatomical evidence for control over the vocal apparatus; the second is the estimated social time demands across hominin evolutionary history; and the third is the pattern of evolving mentalizing skills. None of these could reasonably be said to constitute infallible evidence on their own, but taken together they do seem to converge on a common position.

The neuroanatomical evidence comes in two forms, both based on the observation that modern humans differ from great apes and other primates. One is the size of the spinal cord in the thoracic region in the chest (the point at which the nerves that control the diaphragm and chest wall muscles exit the spinal column) and the other is the size of the hypoglossal canal in the base of the skull (the aperture through which cranial nerve XII, which innervates the tongue and mouth, passes). Both show significant enlargement, relative to body size, in modern humans compared to great apes and monkeys, and this enlargement is plausibly associated with speech (control of the diaphragm to allow the long steady exhalations needed for speech in the first case, and control of the articulatory space in the second). If we could pinpoint the timing of these enlargements in the fossil record, it might indicate at least the point of origin of control over the articulatory space (i.e. speech), even if not language. Although the fossil data are somewhat patchy, taken together they suggest that the australopithecines and early *Homo* all had ape-like values on the two sets of nerves, whereas *heidelbergs*, Neanderthals and AMH all had human-like values.[7]

Two other bits of anatomical evidence have emerged in recent years. One is the position of the hyoid bone. This rather delicate bone acts as a link between the top of the larynx and the base of the tongue: it sits high in the throat in chimpanzees but low in humans, and this is thought to be crucial in allowing us to produce some of the sounds needed for human speech (notably vowels). Because it is so small and delicate, it rarely survives in fossils. However, one Neanderthal hyoid bone was found *in situ* at the Kebara cave site in Israel, and its position is low, as in modern humans. Since then, it has been possible to determine the position of a handful of other hyoids. Those from early *Homo* seem to be as high as in apes. The second piece of anatomical evidence concerns the ear canals – the bony casing within the skull that houses the semi-circular canals that allow us to hear sounds. It turns out that human and chimpanzee ear canals differ in important ways that affect the ability to detect human speech, and recent CT scans of archaic humans from the Sima de los Huesos site in northern Spain (dated to around 500,000 years ago) show that these people had human-like canals.

Taken together, these four kinds of anatomical evidence would seem to suggest that at least the capacity for speech, even if not language, evolved with archaic humans, sometime around 500,000 years ago. Fig. 7.3 plots these data against the predicted social time requirements for the various hominin species. Notice that the anatomical evidence for vocal control seems to appear just where the social time requirement breaches, for the first time, the 20 per cent barrier. It is here, if anywhere, that the demands of social bonding become most challenging because no species of primate devotes more than 20 per cent of their day to social interaction. So it should perhaps be no surprise to find that the anatomy for vocal control undergoes a phase shift at the same time.

However, we need to draw a clear distinction between speech (as a form of vocalization) and language (as grammatically ordered utterances). These anatomical data tell us only that archaic and modern humans had sufficient vocal control to allow complex vocalizations: whether or not this involved *language* is an entirely different matter. I suggested in Chapter 6 that wordless humming or singing might well have evolved among the archaic humans. Singing uses exactly the same vocal apparatus and requires the same vocal control as speech; it does not always require language even though we often sing words.

In short, the anatomical data probably tell us when vocal control emerged, but not when language evolved. Of course,

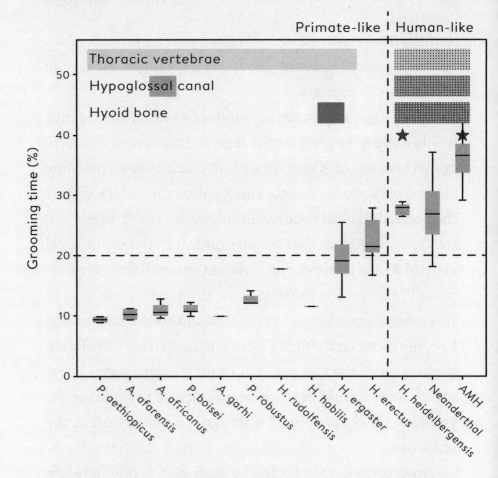

Figure 7.3

Estimating when speech evolved. Median (with 50 and 95 per cent range) grooming time for the main hominin species, predicted using the grooming time regression equation from Fig. 2.1 with the community size data from Fig. 3.3. At the top of the graph are shown the distributions of primate-like (plain shades) and modern human-like (hatched shades) thoracic vertebrae, hypoglossal canals and hyoid bones. The ★ symbol denotes ear canals capable of hearing human speech sounds. Human-like forms of all of these anatomical markers of speech production and perception seem to appear for the first time around 500,000 years ago in archaic humans (*Homo heidelbergensis*). (Note that this graph differs from Fig. 5.3a in that the values for *heidelbergs* and Neanderthals have been corrected for latitudinal effects on the visual system, and hence frontal neocortex volume, following Pearce et al. 2013.)

increased vocal control and a more complex vocal repertoire are important precursors for the evolution of language, and it is very likely that a more complex vocal repertoire evolved quite early on in hominin evolution in response to increasing group size. Indeed, we see exactly this kind of response to increasing group size in both birds and monkeys. In a particularly neat set of field and laboratory experiments, Todd Freeberg showed that the calls of chickadees (a small North American bird of the tit family) become more complex as their group size increases on a day-by-day basis. More interestingly, the American anthropologist Seth Dobson showed that primate facial and gestural repertoires become more complex (and the brain regions that process these become larger) as group size increases, while the British biologists Karen McComb and Stuart Semple showed that the same was true for primate vocal repertoires. Thus the *capacity* to increase the vocal (and gestural) complexity of communication in response to increasing group size is of very ancient origin, and not at all something specific to humans – never mind being dependent on language.

The third source of data concerns mentalizing competences. Mentalizing is crucial for language because both the speaker and listener have to work hard at understanding each other's intentions. The speaker has to intend that the listener understands just what it is that the speaker is trying to persuade the listener to believe, and the listener has to do the same in reverse (Fig. 2.2). The listener might be able to get away with second order intentionality, but the speaker probably requires third order – and that's before they even start talking about someone else. There is a second reason why mentalizing competences are important for language, and that is that the reflexive structure of mentalizing bears an uncanny resemblance to the embeddedness of clauses in the grammatical structure of sentences. The limits on both appear to be at fifth order. Unpacking meaning in language-as-we-know-it can only be achieved by individuals who can cope with fifth order intentionality.

We know from our neuroimaging experiments that mentalizing competences correlate with the volume of the mentalizing network in the brain, and especially with the volume of the orbitofrontal cortex (see Chapter 2), and this provides important support for the claim that, across primates, mentalizing competences correlate with frontal lobe volume (Fig. 2.4). Given this, we can use the equation for Fig. 2.4 to estimate the mentalizing competences of fossil hominins, since they must, by definition, be strung out between the great apes and modern humans. Fig. 7.4 plots the results. As a group, the australopithecines cluster nicely around second order intentionality, along with the other great apes; early *Homo* populations all sit at third order intentionality, while archaic humans and Neanderthals can just about manage fourth order; only fossil AMH (like their living descendants) achieve fifth order.

What this tells us is that even if the Neanderthals did have language (as many would like to claim), then we can confidently assert that their language definitely wasn't as complex as that which we find in modern humans. Given the crucial role of mentalizing skills in language, the difference in how the two species' brains were organized must inevitably have meant that the quality of Neanderthal language was very

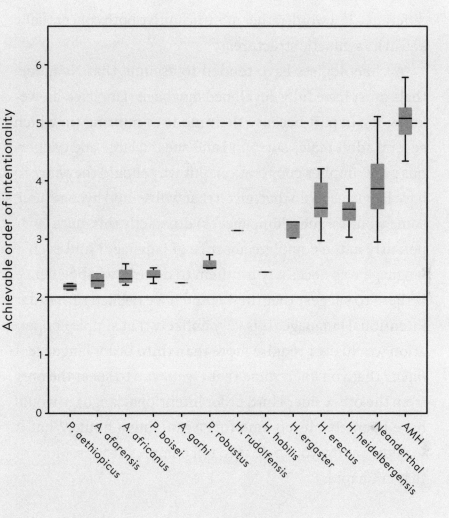

Figure 7.4

Achievable levels of mentalizing in the main hominin species. The plotted values are the median (with 50 and 95 per cent range) order of intentionality predicted for individual fossil specimens by interpolating cranial volume (transposed into equivalent frontal lobe volume) into the equation for Fig. 2.4. The solid line marks the upper limit for great apes (second order), and the dashed line marks the norm for modern human adults (fifth order). Cranial volumes of archaic and Neanderthal specimens are corrected for the effect of latitude on the visual system (following Pearce et al. 2013); *H. erectus* cranial volumes are uncorrected.

different – in a word, rather more primitive both conceptually and in its syntactic structure.

Archaeologists have tended to assume that Neanderthals must have fully developed language (language-as-we-know-it), in part because their mode of hunting (in which several adult males surround and spear a large and dangerous prey) implies cooperation. But why should they *have* to have language to do that, given that wolves and hyenas (and, some would argue, chimpanzees) do exactly this quite competently in the complete absence of language? And even if language was necessary for them to do this, what evidence is there to suggest that they would have needed fifth order intentional language? It is very unlikely that simple cooperation would ever require more than third order language: I *intend* that you *understand* that I *want* you to spear the prey from the other side. Third order intentional language would have been absolutely fine for organizing a hunt. What it wouldn't allow you to do is to tell complex stories. More of that in Chapter 8.

A small problem of reproduction

Hidden behind these big brains lies an altogether less forgiving problem. In mammals, birth occurs when brain development is more or less complete and the baby can survive on its own. Since neurological tissue can only grow at a constant rate, any species that wants to increase its brain size can only do so by increasing proportionately the duration of gestation to allow more brain growth to take place. To accommodate this, large-brained species like primates have fewer babies

less often, and have to extend their lifespan to allow sufficient time to produce enough babies to ensure the survival of
their lineage.

The size of the human brain has necessitated some
unusual accommodations. Based on the standard mammalian
pattern, modern humans should have a 21-month pregnancy:
this is the point at which human babies reach the same stage
of brain development that the babies of all other primates
achieve when they are born.[8] Unfortunately, thanks to an
accident of very early pre-mammal vertebrate history, babies
have to pass through a hole formed between the two halves of
the pelvis, and the size of this hole places a constraint on the
size of the head that can pass through it. This problem was
exacerbated by the evolution of bipedalism, which resulted
in the pelvic bones being reshaped into a bowl-like form to
support the guts and trunk, thereby narrowing the pelvic
inlet and birth canal – a problem that has become known as
the obstetric dilemma.[9] Of course, the problem could easily
have been solved by evolution creating women with hips wide
enough to pass a 21-month-old foetus with a fully developed
brain. Natural selection could easily have managed that, just
as it has done so for the elephant with its similarly long gestation. However, the fly in the ointment is that the genus *Homo*
was designed for a nomadic lifestyle long before super-big
brains came on the scene: hips a metre or more wide are not
ideal for walking or, more importantly, running. Most of the
gains made in terms of nomadism by *Homo ergaster* would
have been wiped out.

The compromise solution that our ancestors came up
with was to reduce the length of gestation to the absolute

minimum needed to produce a baby that could just survive on its own, and then complete brain growth outside the womb – a kind of kangaroo solution. Ape and monkey babies are able to totter about on their own within hours of being born, but human babies do not achieve this stage until they are about 12 months old. This is why modern human babies that are born prematurely (i.e. at less than about 7 months' gestation) are so much at risk and require all the skills and technology of modern medicine to enable them to survive. In the past, most of these would have died soon after birth.

Even so, the birth process is a far more difficult business for modern humans than it is for any monkey or ape, and nature has had to engineer some eye-watering solutions for getting an excessively large-headed baby through an unnecessarily small hole. The cartilages that join the two halves of the mother's pelvis soften during labour and allow the pelvic bones to move apart as the baby forces its way through – which is why women's hips never quite return to their original size after birth. In addition, the bony plates that make up the baby's skull are separated at birth (they don't fuse to make the skull rigid until brain growth has ceased around 5–7 years of age); under pressure from the birth canal, the edges of these plates slide ever so slightly over each other, compressing the baby's head just enough to help it slip through.

We know that this change in reproductive style, with the associated extension of the period of childhood, came about quite late in hominin evolution because we can determine how long it took children to reach adulthood by counting the perikymata[10] in the tooth enamel of fossil skulls. The Nariokotome Boy was nearly full grown even though he was only about 8 years old (although some estimates suggest an age nearer to 12) at the time he died. At this age, he was already 1.5 m tall. A modern human child would not have reached his stage of development until many years later.

On balance, it is very unlikely that the modern human pattern of development, with its premature birth and an extended immaturity, evolved much before the appearance of archaic humans (*Homo heidelbergensis*). It is possible that *heidelberg* brains were just small enough to squeeze through without too much trouble, but the Neanderthals were definitely another matter. By the end of their time, their brains were as large as ours: they would have faced the same obstetric dilemma as we do. So the bottom line is that the switch to a shortened human-like gestation might have evolved independently in both the Neanderthal and modern human lineages – or evolved in their common ancestor. But it certainly didn't evolve before.

However, the evolution of a greatly extended *childhood* seems to have occurred only in the modern human lineage. A recent study of Neanderthal perikymata suggests that they developed significantly faster after birth than modern humans do, and so reached puberty and adulthood several years earlier. This isn't too surprising. In primates, the best predictor of nonvisual neocortex volume is the length of the socialization period (the time between weaning and puberty), reflecting the fact that, once the brain has stopped growing, large neocortices need disproportionately long periods of socialization during which the juvenile learns the social skills on which its future life will depend. Since, as we saw in the last chapter, Neanderthals had much smaller frontal and temporal lobes than modern humans and would thus have been socially less complex, it is hardly surprising that they could get away with a shorter period of socialization. This would have had very significant consequences for their ability to develop a sophisticated culture because of the much shortened period during which social learning could occur.

The problem of supporting a large brain is so demanding that it may have resulted in the rather intriguing possibility that we used external help to do so in the form of the tuberculosis bacterium. Although TB is often seen as a terrible disease, in fact only 5 per cent of those who carry the bacterium are symptomatic, and only a proportion of those die (usually when the symptoms are exacerbated by poor living conditions). In fact, the TB bacterium behaves much more like a symbiont than a pathogen[11] – even though, like many of our other symbionts, it can become pathogenic under extreme conditions. The important issue is that the bacterium excretes nicotinamide (vitamin B_3), a vitamin that turns out to be crucial for normal brain development. Chronic shortage of B_3 rapidly triggers degenerative brain conditions like pellagra. The crucial point here is that vitamin B_3 is primarily available only from meat, and so a supplementary source of B_3 might have become desirable once meat came to play a central role in our diet. Hunting, unlike gathering, is always a bit chancy, and meat supplies are invariably rather unpredictable. This may have become even more crucial during the Neolithic: cereals, in particular, are poor in vitamin B_3 and a regular alternative supply might have become essential after the switch to settled agriculture.

Although it was once thought that humans caught TB from their cattle after domestication around 8,000 years ago, the genetic evidence now suggests that human and bovine TB are completely separate strains, and that the human form dates back at least 70,000 years. If so, its appearance is suspiciously close to the sudden upsurge in brain size in anatomically modern humans that started around 100,000 years ago.

Whatever happened to the Neanderthals?

Having survived successfully in Europe and western Asia for the better part of 300,000 years, the Neanderthals disappeared sometime around 28,000 years ago (the last date for which we have any fossils). So why did this incredibly successful species go extinct, leaving the European stage exclusively to their AMH cousins from Africa? There have probably been more explanations offered for this iconic event than for any other in human evolutionary history. These include: inability to cope with the intense cold of the last Ice Age; a major series of volcanic eruptions in the Caucasus that created nuclear-winter conditions;[12] a lack of game to hunt as the ice front (or the volcanic eruptions) pushed the great herds southwards faster than the Neanderthals could follow them; ecological competition from modern humans; populations that were so small and scattered they were unable to invent or maintain cultural innovations; they were wiped out by new diseases that modern humans brought with them from Asia or Africa; they were absorbed into the modern human population by interbreeding; and, perhaps inevitably, modern humans deliberately killed them off in the mother of all racial-cleansing events. There is probably a grain of truth in all of these, but the real answer is almost certainly that they were caught by an unfortunate combination of circumstances.

Like contemporary modern humans, Neanderthals were forced southwards by the advancing ice sheet that, at its peaks, repeatedly pushed them into the pockets in the underbelly of Europe (Spain, Italy, the Balkans). One of the classic findings of conservation biology is that, when a species is forced into relict habitats like this, their reduced and isolated populations become much more susceptible to random local extinctions that could not be replaced by immigration from elsewhere. This effect would have been exacerbated for the Neanderthals by the smaller sizes of their communities (see Chapter 6).

Had they had to cope with only one of these factors, the Neanderthals might well have survived. But all of them together combined to overwhelm them. It probably didn't help that any contacts they had with modern humans would have exposed them to novel pathogens to which they had no immunity – much as, nearly 30,000 years later, the Amerindian populations collapsed in the face of relatively innocuous diseases that European colonists brought with them from the Old World.

And then there is the thorny question of Neanderthal material culture – the tools and clothing that they made. In recent years, there have been repeated attempts to claim that Neanderthal material culture was a match for any of that produced by modern humans – or, at least, that it was showing signs of heading in that direction just before they went extinct.[13] But the bald fact is that it really wasn't in the same league: Neanderthal tools lacked the diversity, the creativeness and the fineness of production that characterized contemporary modern human material culture. By and large, their tools were more explicitly functional and lacked the frivolousness of much later AMH artwork. Given the differences in mentalizing competences, this is hardly a surprise: lacking one whole level of intentionality would have limited their ability to imagine objects in a piece of raw stone or ivory, or to be as technically inventive.

One likely suggestion, then, is that AMH were able to avoid the Neanderthals' fate under the prevailing conditions of climatic stress mainly because their bigger functional brains allowed them to have larger trading networks *and* to be more culturally creative. In Chapter 6, we noted that not only were Neanderthal community sizes much smaller than those of contemporary modern humans (about two-thirds the size: Fig. 6.5), but the distances over which they traded or exchanged raw materials was an order of magnitude smaller: 70 per cent of the raw material of tools found at Neanderthal sites had travelled *less* than 25 km, whereas 60 per cent of those from contemporary modern human sites had travelled *more* than 25 km and some had travelled as much as 200 km. Larger social networks over very much larger geographical areas would have buffered modern humans against local extinctions by allowing them to seek refuge with friends in a way that would not have been possible for Neanderthals. And if modern humans had already added the outer network layers (the 500 and 1,500 layers: Fig. 3.4) on to their network structure before they left Africa, as seems likely, this would have exaggerated the magnitude of this difference considerably.

Differences in brain organization between modern humans and Neanderthals may have implications for cultural complexity quite separately from mentalizing abilities. In primates, the ability to project behavioural events into the future and plan ahead is determined by the same regions of the cortex (the very front part of the brain) that are implicated in mentalizing. Across anthropoid primate evolution, and into modern humans, it is these prefrontal regions that have progressively increased in size.[14] Neanderthals, with their smaller prefrontal cortices, would have had much reduced planning capacities. And that in turn may well have implications not just for their ability to design tools and other manufactured objects, but also for their ability to predict the future consequences of their behaviour, not to mention their ability to inhibit their responses to a situation.

One aspect of cultural adaptability that almost certainly made the difference for modern humans is clothing. Leslie Aiello and Peter Wheeler calculated the thermal properties of clothing that Neanderthal and modern humans would have needed to survive at their various European sites during the last Ice Age, based on the temperatures estimated from the Late Palaeolithic Stage 3 climate models. Even though Neanderthal living sites were, at any given time, always south of those used by contemporary modern humans, their physical adaptations to cold climates would have allowed them to survive with clothing that was much less warm than modern humans would have needed. They were much better thermally adapted than modern humans were, or indeed are.

My guess is that this reflects the difference between seamed clothing (providing a tight, thermally efficient fit to arms and legs) and crude wrappings. It is surely no accident that we begin to find eyed needles (usually of bone) at fossil sites associated with modern humans from about 30,000 years ago in Europe; in fact, there is a long history of boring tools associated with modern humans, since they were using awls for boring shells to make necklaces from at least 100,000 years ago in southern Africa. There is no evidence that Neanderthals had these kinds of tools for fine quality work. This almost certainly reflects the differences in mentalizing abilities (or at least the cognitive skills that underpin mentalizing) between the two species.

Rather more spectacular evidence of clothing comes from the Upper Palaeolithic burials of modern humans. At Sungir, on the upper reaches of the River Volga just northeast of Moscow, two children were buried head to head in a double grave dated to around 22,000 years ago. What is remarkable about this burial is the presence of very large numbers of pierced beads, which would have required many thousands of hours of skilled work to shape and bore. One child (possibly a male) had 4,903 beads lying as though they had been attached to tight-fitting clothes. There were also 250 pierced arctic fox teeth around his waist that probably formed part of a belt, and an ivory pin at his throat that might have secured a cloak. The other skeleton (possibly a female) had a total of 5,374 beads that seem to have been attached to her clothing, as well as an ivory pin at her throat. Modern humans must have been using clothing for some considerable time prior to this, because beads and buttons made from ivory, bone, amber, shell and stone have been found at many European Upper Palaeolithic sites from around 35,000 years ago, and considerably earlier in Africa.

Some rather surprising evidence for clothing is provided by molecular genetics. Modern humans possess two kinds of lice that belong to separate subspecies: the head louse *Pediculus humanus capitis* and the body louse *P. h. corporis*. The two subspecies do not interbreed because they occupy different habitats on the body: one is confined to hair on the head, whereas the other lives in our clothing (i.e. on the trunk). The fact that the latter can only survive when clothing is available for it to hide in suggests that it evolved after humans began wearing clothes on a regular basis. Comparison of the mtDNA of these two subspecies from 12 different parts of the world shows that body lice evolved from the head louse, and that the two subspecies had a common ancestor sometime around 100,000 years ago – indicating that this was when body clothing started to be worn on a regular basis. Significantly, this precedes the point at which anatomically modern humans first emerged out of Africa and began to colonize Eurasia, suggesting that when modern humans arrived in Europe 40,000 years ago they were already kitted out in close-fitting clothes that they had developed in Africa tens of thousands of years previously. It also coincides with the final phase of brain growth.

The inevitable question that always hovers in the background, however, is whether the Neanderthals went extinct because anatomically modern humans killed them off. We will never know for sure, of course, but at least some of the many injuries on Neanderthal skeletons are probably not due just to carelessness when hunting. Careful analysis of the injuries on the 36,000-year-old St Césaire juvenile Neanderthal suggests that some were due to a sharp instrument and thus probably the consequence of violence. Whether this was violence between Neanderthals or between AMH and Neanderthals we will never know.

The possibility that modern humans interbred with the Neanderthals has generated considerable media as well as academic excitement. The original claim was based on what seemed to be a mixture of Neanderthal and modern cranial traits in an AMH infant from Gibraltar, but, in reality, anything based on very young individuals is at best speculative. More serious evidence has come from the finding that 2–4 per cent of modern European DNA is shared with the Neanderthals (and not with modern Africans). Even more interesting has been the suggestion of interbreeding between AMH and the Denisovans in the Far East. Genetic sequencing of the Denisovan genome tells us that about 4–6 per cent of the DNA of modern Melanesians and Australian Aboriginals might be of Denisovan origin. In neither case, of course, does this indicate that interbreeding was exactly a regular affair – and, indeed, some have even suggested that it may simply reflect genetic convergence when occupying similar habitats and encountering similar selection pressures.

The fact that humans were interbreeding with archaics does not mean that the two species were engaged in friendly exchanges of brides, of course. Modern human colonizers have considerable historical form in stealing women from native populations. This is known to have happened repeatedly (in the Americas and India, in particular), but there is also good genetic evidence for this happening much deeper in history. In southern England, for example, women have mainly Celtic mitochondrial DNA, but the men's Y-chromosomes exhibit a very distinct gradient from being mainly Anglo-Saxon in the east to mainly Celtic in the west, suggesting that during the fifth- and sixth-century Anglo-Saxon invasions from the Continent, the male invaders were availing themselves of the local women and, one way or another, preventing the local British males from breeding. A similar story emerges from the genetics of Iceland: 80 per cent of Icelandic male Y--chromosomes are Norse in origin, but 63 per cent of the women's mitochondrial DNA is Celtic, presumably reflecting the fact that the women had been carried off, forcibly or otherwise, by Viking males on their way to Iceland.[15] Similarly, 7 per cent of all male Y-chromosomes in Asia and the eastern fringes of Europe are Mongol in origin, and are thought to derive from the activities of Genghis Khan and his male relatives (in particular) during the surprisingly short-lived Mongol conquests of the thirteenth century.[16]

In other words, if historical human behaviour is anything to go by, the modest amount of Neanderthal DNA in the modern human genome probably reflects a certain amount of pillage by unpaired AMH males. If there really was interbreeding between archaic and modern humans during the dying days of the Neanderthals and Denisovans, then it likely wasn't something that the archaics were entirely happy with, and that in turn is likely to have resulted in some resistance on their part. If so, then the seemingly modest differences in community size between the two species are likely to have proved decisive. Modern humans would have been able to draw on much larger numbers of males from communities that likely had significantly greater social cohesion. Moreover, modern humans would have been able to marshal allies from a much wider geographical area than the archaics could.

How Kinship, Language and Culture Came to Be

The appearance of anatomically modern humans in Europe around 40,000 years ago is marked by a new and distinctive array of tools and other artefacts often referred to as the Upper Palaeolithic Revolution. Novel and more powerful weapons become widespread, including technological advances such as javelin-like throwing spears, spear-throwers,[1] and bows and arrows. There are fine awls and needles for working clothes, oil lamps and huts with cobbled floors. This dramatic burst of cultural activity even spills over into decorative art – cave paintings, Venus figurines carved from ivory and stone (Fig. 8.1), or horse-headed handles for spear-throwers. There are bone flutes and calendars carved on ivory plaques. There are deliberate burials where the body is laid to rest with a rich array of grave goods, speaking perhaps of a sense of an afterlife where the dead will need the trappings of everyday life. The modern mind had finally arrived.

This does not, of course, represent a moment when the light bulb went on in somebody's head and all of a sudden he or she was inventing all kinds of novel artefacts. There had been a long slow build-up to this that can be documented in the archaeology of Africa for at least 100,000 years. The

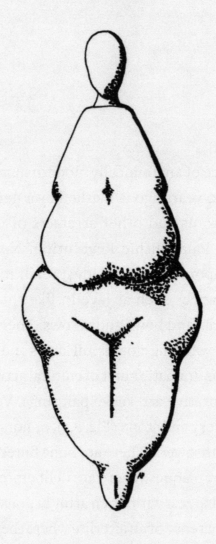

Figure 8.1

The Venus of Lespugue, in southern France, dated to 25,000 years ago. Redrawn from Lewis-Williams (2002). © 2014 Arran Dunbar.

reality, however, is that the period starting from around 40,000 years ago, and building exponentially through the next 20,000 years, is marked by an extraordinary explosion in the quantity, quality and diversity of artefacts, especially but by no means exclusively in Europe.

Although many of these cultural developments involve more sophisticated tools, and so bear on the business of food-finding, many do not. It would be hard to justify the cave art and the Venus figurines in these terms. Rather, they may have been part of an attempt to solve a critical problem that had bedevilled hominin evolution since its earliest days, but which reached positively crescendo proportions with the appearance of *Homo sapiens* – how to maintain social coherence in very large social groups, especially when these groups were distributed over a wide area. And since these mechanisms are all essentially cultural, language must inevitably have played a crucial role.

Why language evolved

Historically, the consensus has been that language evolved to allow humans to exchange factual information about the physical world, but an alternative view is that language evolved, in modern humans at least, to facilitate social bonding. Both views, of course, agree that the grammatical structure of language is designed to allow the exchange of information. They differ in which facts are considered most important to our survival (instrumental versus social knowledge), and therefore on the primary function of language, with the alternative function being seen as a trivial by-product – something that comes for free once you have language for its main function.

The problem is that there is no principled reason for preferring one view over the other, and testing between these options is difficult in the absence of comparative data (we are, after all, the only living species with language) or relevant archaeological evidence (alas, conversations don't fossilize). One possible way to distinguish between the two hypotheses is to ask whether people find it easier to remember one kind of information than the other, assuming that the human mind is designed to pay particular attention to some kinds of information (the kind for which language originally evolved) and not others. The logic of this is simple: a trait's original design features will continue to be the most natural for it *even if* it was subsequently adapted to serve other functions in addition. Of course, this is not a perfect test of the historical sequence in which these traits were acquired, but it is probably the best we can do.

Three different versions of the social hypothesis have been proposed: language for the exchange of information about social relationships (my original gossip hypothesis), language for making formal arrangements and public declarations (the social contract hypothesis, originally proposed by Terry Deacon) and language for mate attraction and retention (the Scheherazade hypothesis, originally proposed by the evolutionary psychologist Geoffrey Miller).

The gossip hypothesis is very simple. It suggests that language evolved to allow the exchange of information that could be used to create and foster social relationships, enabling individuals to maintain a level of knowledge about others in large, dispersed networks that would be simply impossible if this had to be done only by face-to-face interaction. In other words, we can exchange knowledge about who is doing what with whom in a way that direct observation would not allow.

In proposing the social contract hypothesis, Deacon had marriage arrangements in mind. He pointed out that, among hunter-gatherers, the sexual division of labour in which the men go off hunting for several days at a time while their mates are left behind unguarded and exposed to rivals who might be tempted to interfere with them is very risky; some common agreement as to who is spoken for and by whom is therefore necessary to guarantee the paternity of one's children, and this, he suggested, came to form the basis of symbolic arrangements like marriage. It is a plausible argument, since paternity certainty is a major evolutionary issue for males: women always know who their children are, but men can never be 100 per cent certain and they risk unwittingly investing in other men's children if they are wrong – a form of genetic altruism that natural selection will inexorably select against. But are these kinds of contracts the cause or consequence of language? Contracts require the prior existence of a division of labour, and that probably didn't happen until males began hunting big time.

In contrast, Miller argued that language evolved through sexual selection as a form of advertising to prospective mates, and as a mechanism for keeping mates engaged and interested after pairbonding. This, he argued, is why language (and men's language in particular) is so often flamboyant and redundant: being able to play with language shows how smart you are, and hence what good (brain) genes you have. Sexual selection is an extremely powerful force in evolution, and it is exceptionally good at exploiting windows of opportunity offered by natural selection. Hence, the use of language for advertising could very easily be a consequence of sexual selection exploiting language rather than its primary driver.

We have conducted two sets of experiments to test between these functions of language, using memory for the contents of short stories as the basis for deciding which type of information the human mind is best adapted for. In both cases, people remember social information much better than they remember physical world factual content. This also turns out to be true of the content of microblogs on the internet. Between them, these studies suggest that social information exchange has priority for us, at least in terms of the kinds of information that we pay attention to and remember most easily.

The second of our studies, designed and carried out by Gina Redhead, explicitly sought to test directly between the four hypotheses (three social and one instrumental). Each of the hypotheses was represented by a different story, with an additional social version that focused on romantic relationships (a pure gossip version of the social information exchange hypothesis). The neatness of the experimental design was that Miller's Scheherazade hypothesis was run as an instrumental (rather than social) story couched in flamboyant language. People remembered the three social stories significantly better than the two instrumental ones (Fig. 8.2), but the social versions did not differ among themselves. More importantly, the flamboyant version of the

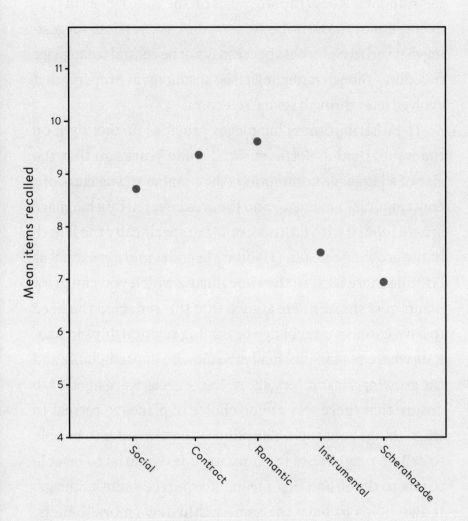

Figure 8.2

Mean number of items remembered from short vignettes representing alternative hypotheses for the function of language. From left to right, these represent alternative versions of the gossip hypothesis (general social gossip and gossip about romantic relationships), Deacon's social contract hypothesis, the instrumental hypothesis (factual information exchange) and Miller's Scheherazade hypothesis (flamboyant language, in this case applied to the instrumental story).

Redrawn from Redhead and Dunbar (2013).

instrumental story (the story was on the role of bees in producing honey) fared no better than the basic version, suggesting that verbal cleverness per se may not be central to language function – though it might well be an emergent property that evolved later through sexual selection.

The distribution of languages provides further light on the issue. Daniel Nettle showed some years ago that the size of a language community (the number of speakers of a contemporary language) and the area covered by a language are correlated with latitude, or more specifically the length of the growing season. (Habitats become more seasonal as latitude increases, so the time during which you can grow plants gets shorter.) He argued that this reflected the need to have extensive exchange or trading relationships in habitats where seasonality made the climate unpredictable and the growing season very short. Large areas were needed to ensure that there was a wide choice of places to retreat to when things got tough. The point is that, in order to be able to call on neighbours for support, it is essential to be able to talk to them, and that means sharing the same language. It also helps to have the same worldview (moral beliefs, understanding of the world, etc.), and shared worldviews come through having the same language. Indeed, we found in our studies of friendships that shared language and shared worldview play a seminal role in the strength of everyday friendships (see Chapter 9).

Aside from the fact that our experiments suggest that humans don't seem to remember facts about the non-social world terribly well, it is clear that we aren't all that keen to pass on information about the physical world unless it benefits our close family and friends. This was demonstrated in a study of lobster fishermen's radio exchanges about the location of lobster swarms off the Maine coast in the northeast of the USA. Craig Palmer found that in large communities that contained many strangers, fishermen were significantly less willing to broadcast the location of rich lobster pickings when they found them compared with those from small communities where almost everyone was part of the same social network. Information exchange is only something we engage in *once* we have used language to establish a relationship.

Language has the obvious advantage that shifting some aspects of social bonding from a physical channel (grooming) to a vocal channel may allow several individuals to be 'groomed' simultaneously, allowing us to build a larger community. Language might do this in three quite different ways. One is by telling each other how we see the world (creating a common worldview); a second is through story-telling (stories about who we are and where we have come from); and the third is through making people laugh by telling jokes. I will say more about the first of these in the next chapter, so here I will focus just on the other two. The evolution of language changed the nature of laughter for ever. Up until this point, laughter was a form of chorusing in which – most likely – spontaneous bursts of laughter were triggered by some event (perhaps play, perhaps another's misfortune) in what may amount to a simple commentary on the event in question. However, these kinds of events occur only sporadically and unpredictably, and so laughter choruses could only have occurred at very irregular intervals. Language changed all that by allowing us to manage laughter much more effectively: we could trigger choruses of laughter at any time, day or night, by telling jokes. Jokes are almost always about other people's mind states, and so require high order intentionality to appreciate them, particularly if they depend on metaphor (which is impossible to understand without theory of mind). So language probably ramped up the effectiveness of laughter as a bonding mechanism in a way that had not previously been possible. However, jokes are really a by-product of having language rather than being the driver for language itself. Creating a joke depends on being able to make statements about other people's mind states before we can make fun of them.

Telling a story – whether about what happened in history, or about the ancestors or who we are and where we came from, or about the people that live beyond the horizon, or the inhabitants of a spirit world that none of us can experience directly – creates a sense of community by binding us into a network of people who share a common view of the world. Importantly, it may even allow us to appreciate why the folk in the next valley should – or should not – be considered part of our extended community, so potentially creating an additional layer beyond the limits of the 150 that otherwise demarcate the set of individuals of whom we have personal knowledge.

It may even be that, irrespective of their content, telling stories round the hearth may create a sense of bondedness. Roused emotions seem to be very good for creating a sense of community, possibly because they trigger the release of endorphins. Puberty rituals in most traditional societies typically involve frightening experiences, and this seems to be crucial in creating the sense of lifelong brotherhood and mutual commitment that those who go through these rituals together share. It may also be no accident that story-telling is particularly engaging when it is done at night. That might be because fear of the dark naturally allows a skilled story-teller to heighten the audience's emotional response, but it may be the sense of intimacy that the dark creates by shutting out the rest of the world.

In concluding this section, it is worth noting that languages have one very odd quality for a faculty that is supposed to function as a mechanism for facilitating the exchange of instrumental information: they constantly fractionate into dialects which very quickly give rise to new, mutually incomprehensible languages. English now officially consists of six different languages[2] whose origins date back less than 1,000 years, and in some cases only a few hundred. If language is designed to facilitate cooperation, why would the language faculty be *so* ineffective as to make it difficult for neighbouring groups of individuals to understand each other? In short, why should dialects provide such obvious markers of where you come from?

The answer seems to be that a dialect identifies a small community of people who come from the same place – and, hence, at least in small-scale societies, are likely to be related to each other. Even in the contemporary world, dialects change so fast (on the scale of generations, if not decades) that they allow us to identify not only where a person was born but also which generation they belong to. This doesn't make any sense at all if language exists to allow the exchange of instrumental information, but it makes perfect sense if language evolved to create and bond small, exclusive communities.

Naming kin

There is one important aspect of language that some have argued constitutes the origin of language itself – the naming of kin. There is no particular reason to assume that ability to name kin relationships was in any way ancestral, although it may well be the case that naming individuals appeared very early. On the other hand, labelling kinship categories (brother, sister, grandfather, aunt, cousin) is quite sophisticated: it requires us to make generalizations and create linguistic categories. And it probably requires us to be able to handle embeddedness, since kinship pedigrees are naturally embedded structures.

Kinship labels allow us to sum up in a single word the exact relationship between two individuals. The consensus among anthropologists is that there are only about six major types of kinship naming systems – usually referred to as Hawaiian, Eskimo, Sudanese, Crow, Omaha and Iroquois after the eponymous tribes that have these different kinship naming systems. They differ mainly in terms of whether they distinguish parallel from cross cousins[3] and whether descent is reckoned unilaterally or bilaterally.[4] The reasons why these naming systems differ have yet to be explained satisfactorily. Nonetheless, given that one of their important functions is to specify who can marry whom, it is likely that they reflect local variations in mating and inheritance patterns. The Crow and Omaha kinship naming systems, for example, are mirror images of each and seem to be a consequence of differing levels of paternity certainty (as a result, one society is patrilineal, the other matrilineal[5]). Some of these may be accidents of cultural history, while others may be due to the exigencies of the local ecology. Kinship naming systems are especially important, for example, when there are monopolizable resources like land that can be passed on from one generation to the next and it becomes crucial to know just who is entitled, by descent, to inherit.

Anthropologists have sometimes claimed that biology cannot explain human kinship naming systems because many societies classify biologically unrelated individuals as kin. This is a specious argument for two separate reasons. One is that the claim is based on a naïve understanding of what biological kinship is all about. This is well illustrated by how we treat in-laws. In English, we classify in-laws (who are biologically unrelated to us) using the same kin terms that we use for real biological relatives (father-in-law, sister-in-law, etc.). However, as Max Burton-Chellew and I showed, we actually treat them, in emotional terms, as though they were real biological kin, and we do so for a very good biological reason: they share with us a common genetic interest in the next generation. We tend to think of genetic relatedness as reflecting past history (i.e. how two people are related in a pedigree that plots descent from some common ancestor back in time). But in fact, biologically speaking, this isn't really the issue, although it is a convenient approximation for deciding who is related to whom. In an exceptionally insightful but rarely appreciated book (mainly because it is *very* heavy on maths), Austen Hughes showed that the real issue in kinship is not relatedness back in time but relatedness to *future* offspring. In-laws have just as much stake in the off-spring of a marriage as any other relative, and hence should be treated as though they are biological relatives. Hughes showed that this more sophisticated interpretation of biological relatedness readily explains a large number of ethnographic examples of kinship naming and co-residence that anthropologists have viewed as biologically inexplicable.

The second point is that in traditional small-scale societies everyone in the community *is* kin, whether by descent or by marriage; those few who aren't soon become so by marrying someone or by being given some appropriate status as fictive or adoptive kin. The fact that some people are misclassified as kin or a few strangers are granted fictional kinship status is not evidence that kinship naming systems do not follow biological principles: a handful of exceptions won't negate the underlying evolutionary processes associated with biological kinship,[6] not least because everything in biology is statistical rather than absolute. One would need to show that a significant proportion of naming categories cross meaningful biological boundaries, but in fact they never do. Adopted children can come to see their adoptive parents as their real parents, but adoption itself is quite rare; moreover, when it does occur in traditional societies it typically involves adoption by relatives (as anthropological studies have demonstrated). A real sense of bonding usually happens only when the child is very young (and even then the effect is much stronger for the child than for the parents – who, after all, know the child is not theirs).

Given that kinship naming systems seem to broadly

follow biological categories of relatedness, a natural assumption is that they arise from biological kin selection theory.[7] In this respect, kinship does seem to be a very different kind of relationship to friendship. There have been many studies by us and others that have compared people's willingness to behave altruistically towards family and friends at different removes (i.e. close versus distant friends and family); all show that people typically behave more altruistically towards family than they do towards friends, even when they are matched for social distance (i.e. which layer of the social network they fall into). It seems we have a gut response to help relatives preferentially, presumably as a consequence of kin selection.

Nonetheless, some aspects of kinship are genuinely puzzling. One might suppose that kinship is simply a function of having grown up with someone, and so being deeply familiar with them from early childhood – kinship as some kind of intense friendship, perhaps. But some of the more distant categories of kin (second and third cousins, and cousins once removed, as well as great-grandparents and great-great-grandparents) attract almost as strong a response from us as close kin. Yet these distant relationships are purely linguistic categories that someone has labelled for us ('Jack is your second cousin – you share a great-grandmother'). The moment you are told that somebody is related to you, albeit distantly, it seems to place them in a very different category from mere friends, even if you have never met them before. It is remarkable that a linguistic label can trigger an emotional response that is normally the privilege of close kinship based on deep emotional and biological ties.

This raises the possibility that kinship is a short cut that allows us to deal with the complexities of relationships without having to waste a lot of time when making decisions on how to behave towards someone. We only need to know one thing about kin – that they are related to us (and maybe exactly how closely they are related) – whereas with a friend we have to track back through all the past interactions to decide how they actually behaved on different occasions. Because less processing has to be done, decisions about kin should be done faster and at less cognitive cost than decisions about unrelated individuals. This would imply that, psychologically, kinship is an implicit process (i.e. it is automated), whereas friendship is an explicit process (we have to think about it). In a context where everyone in a community is related to everyone else, as they are in small-scale communities, kinship naming helps to identify members of your community in an efficient 'fast-and-frugal' way.

It may be no coincidence that 150 individuals is almost exactly the number of living descendants (i.e. members of the three currently living generations: grandparents, parents and children) of a single ancestral pair two generations back (i.e. the great-great-grandparents) in a society with exogamy (mates of one sex come from outside the community, while the other sex remains for life in the community into which it was born). This is about as far back as anyone in the community can have personal knowledge about who is whose offspring so as to be able to vouch for how everyone is related to each other. It is striking that no kinship naming system identifies kin beyond this extended pedigree with its natural boundary at the community of 150 individuals.[8] So it seems as though our kinship naming systems may be explicitly designed to keep track of and maintain knowledge about the members of natural human communities.

Fitting religion into the mix

Story-telling forms an essential component of all religions: they all tell stories about long-dead ancestors or the beings that occupy the spirit world; hagiographies of their charismatic founder(s) and saints are often also a central feature. In traditional small-scale societies, religion is something you experience directly, often associated with shamans and trance states (Fig. 8.3). Music and dance play a particularly important role in shamanic practices the world over and are used (sometimes with drugs) to psych up the devotees until they lose full consciousness and enter a trance state. Through trance, they gain entry to a transcendental world through which they journey, usually with ancestors or friendly spirits as guides, and perhaps even face challenges set by evil spirits, much as the hobbits do in *The Lord of the Rings*. Clearly, these travels in the spirit world are actually going on inside the individual's own mind, but they are *so* intense that they seem completely real.

At this point, language is not especially important for religion, because religion is something you experience rather than theologize about in some sophisticated cognitive sense. You might well need to be able to tell each other about your spirit travels and come to some common understanding of what you experience, but you don't need a very complicated

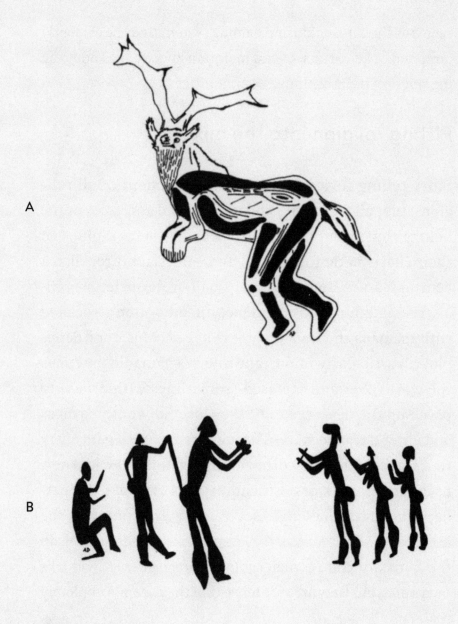

A

B

Figure 8.3

The prehistory of religion. (a) The dancing therianthrope (deer-headed 'sorcerer') from Les Trois Frères gallery, River Volp cave complex, France, dated to 12,000 years ago. (b) San trance dancing: historical San rock art from Southern Africa.

Redrawn from Lewis-Williams (2002). © 2014 Arran Dunbar.

theology for this: there are no gods as such, and most of the creatures that you meet in the spirit world are quite familiar – although some may be mosaics of different beasts, usually referred to as therianthropes (literally, animal humans) (Fig. 8.3a). Trance states and how to enter them are probably very ancient and were most likely discovered by accident, perhaps when music and dance became increasingly important as a bonding mechanism among the *heidelberg* folk and some particularly enthusiastic individuals danced themselves to the point of exhaustion. Trances are not that difficult to engineer once you know how, so it would have been a very small step from their accidental discovery to deliberately inducing them.

Shamanic religions all over the world share a common set of motifs. These include entering the spirit world through a hole or tunnel beyond which there is an explosion of light and a bright world; all agree that travels in the spirit world are fraught with dangers that require a benign guide (either a well-disposed ancestor or a totemic animal), and all fear failure to find the exit hole again (probably a reflection of the fact that trance dancing can be so exhausting that dancers do occasionally collapse and die – or, if you prefer, get left on the other side unable to find the way back).

Trance dances of this kind seem to play an important role in maintaining the social balance within a community. Among the San Bushmen of southern Africa, trance dances are particularly likely to take place when relationships within the extended community have started to unravel as people bicker among themselves. A trance dance restores the equilibrium, almost as though it wipes the slate clean of the toxic memories of the injustices and slights that poisoned relationships. Trance seems to reset relationships back to their default position, allowing the community to function once more as a mutually supportive network of relationships – until the petty slights and injustices build up again over the ensuing weeks and months, and a new trance dance is called for. This probably reflects the fact that trance dancing – if not the trance state itself – is associated with a massive release of endorphins, and it is these, acting in their time-honoured fashion to bond individuals, that make for restored relationships. In addition, endorphins have a widely beneficial effect on our physical and psychological health, and so trance dances likely benefit the health of the community as a whole, as well as its social cohesion.[9]

Religions, then, seem to have evolved as a way of enhancing social cohesion and commitment within very small communities. One of the unfortunate side effects of this is that it inevitably creates an Us-versus-Them, or In-group/ Out-group, mentality. Sharing a worldview, having the same religious experiences and following the same rules of behaviour all add up to drawing a clear-cut division between our community and all those other folk in the next valley who behave badly, do disgusting things and are generally a bad lot.

In a seminal series of papers, the American biologists Cory Fincher and Randy Thornhill showed that the number of adherents to a traditional religion, the size of language communities and the balance between individualism versus collectivism all correlate with latitude: humans have smaller, more inward-looking, strongly bonded communities around the equator, and larger, more outward-looking, individualistic ones nearer the poles. They were able to show that the underlying driver for this relationship is pathogen load: the tropics are a notorious hothouse for diseases, being responsible for spawning a constant stream of new ones even now. They argued that an important way of reducing health risks under conditions of high local pathogen load is to avoid mixing (and especially mating) with other groups: it is best to stick with your own community and your own diseases, since these are the ones to which you have had time to evolve an immunity.

This argument articulates nicely with Daniel Nettle's claims about language communities that I mentioned earlier. Nettle's hypothesis offers a reason why language communities should be large at high latitudes, but doesn't really provide a reason why they should be small in the tropics; on the other hand, Fincher and Thornhill offer a reason why communities should be small in the tropics, but don't provide any principled reason why they should increase at high latitudes. Claiming that communities can become larger at high latitudes because the pathogen selection pressure is greatly reduced (as Fincher and Thornhill do) isn't really an explanation, because there is nothing to override all the social and physiological costs associated with living in large communities that we discussed in Chapter 2. However, combining the two hypotheses provides a satisfying solution: pathogen risk selects for small communities in the tropics, but as this selection pressure relaxes at high latitudes so the need for large-scale trading relationships takes over and forces the evolution of large communities.

We need bigger communities at high latitudes, but we

also need mechanisms to defuse the tensions that inevitably come with them. The rituals of religion seem to play a crucial role in facilitating community cohesion in just this context. This much is suggested by the extent to which, even today, commitment to a religion enhances a sense of belonging to a community and typically makes people more generous towards each other (and occasionally, but by no means always, towards complete strangers as well). The important issue here is not that religion makes people more prosocial (which may well be a beneficial by-product) but that it makes people more committed to the members of their particular community. I return to this issue in the next chapter, since it plays a crucial role in the final transition.

Archaeology and the afterlife

Archaeologists have typically been willing to accept only deliberate burials as evidence for belief in an afterlife, and they have identified deliberate burials as inhumations that are associated with grave goods. Grave goods are taken to imply that the dead need the accoutrements of everyday life in the afterlife. It is difficult to conceive of any reason for taking such care when disposing of the dead if one did not believe in an afterlife. Why not just dump the body out in the woods, or down a sink hole at the back of a cave – perhaps the fate of the many bodies in the Sima de los Huesos?

Deliberate burials are common in the Upper Palaeolithic: more than 100 are known from Europe and western Asia, compared to at most three dozen from the Middle Palaeolithic sites occupied by archaic humans over a period of time many times longer. What marks out the Upper Palaeolithic burials as different from all the earlier ones is the fact that, in most cases, the body was laid on its back, fully stretched out (although two were laid face down, and a few have bent legs). In almost all Middle Palaeolithic 'burials', including those of the Neanderthals, the body seems to have just been dumped.

Many Upper Palaeolithic burials are associated with red ochre, in some cases in sufficient quantities to stain the bones and the surrounding soil. In certain cases, such as the burials at Grotta dei Fanciulli in Italy, the ochre was especially thick on the head and/or the pelvis. Quite why ochre should have been used so liberally in Upper Palaeolithic burials remains puzzling, but it may have been associated with ritual body decoration, as ochre is often used by traditional peoples today as body paint.

Many of these burials are quite elaborate, and are associated with substantial quantities of grave goods. The grave of the two children at Sungir on the Russian steppes, whom we met in Chapter 7, had been dug into the permanently frozen permafrost some 22,000 years ago (which must have taken some considerable effort), and the bodies had been laid to rest with a rich assortment of jewellery, tools, ivory sculptures and decorated antler wands, as well as a human femur (thigh bone) whose shaft had been packed with red ochre. Eleven javelins and a 2.4 m ivory lance had been placed in the grave with them. At the 21,000-year-old Siberian site of Mal'ta, a child was buried along with a diadem, a necklace with pendant, a bracelet, a figurine, buttons, bone points and other tools. In the Cro-Magnon cave shelter in southwestern France, the bodies had been buried with clusters of pierced shells and animal teeth, as were those at the Italian cave site of Grimaldi. At Arene Candide, in Italy, the so-called 'Young Prince' represents one of the most splendid of the Upper Palaeolithic burials. His body was dusted with copious quantities of red ochre and decorated with mammoth ivory pendants, shells that probably once formed a bracelet and several ivory batons; he held a 23 cm flint blade in one hand. In Portugal, the 5-year-old Lapedo child was laid to rest 25,000 years ago with a perforated shell pendant around its neck and a diadem made from the canine teeth of four different red deer. In fact, necklaces and diadems seem to be especially common, having been found in around two-thirds of all Upper Palaeolithic burials.

Upper Palaeolithic burials have another distinctive feature: they often occur in groups. At Sungir, the bodies of the two children were laid head-to-head with a third adult buried nearby, while at Qafzeh, in the Levant, an Upper Palaeolithic woman (Qafzeh IX) was buried with her infant. Five bodies (including an infant) were buried close together in the Cro-Magnon rock shelter in France, while at Dolni Vestonice in the Czech Republic three adults (probably a female and two males) were buried together. The largest multiple burial consists of 18 individuals buried (mostly at different times) at Předmostí in the Czech Republic. Multiple burials of this kind are virtually unknown prior to the Upper Palaeolithic. Such burials suggest a sense that the dead are seen as going together to the same spirit domain or perhaps, in cases where the burials are not contemporary as at Předmostí, as joining those who had predeceased them.

What these findings tell us is that at least by 25–30,000

INTENTION-ALITY LEVEL	POSSIBLE STATEMENTS OF BELIEF	FORM OF RELIGION
1st	I believe that god [. . . exists]	existential belief
2nd	I believe that god is willing [. . . to intervene if you disobey his laws]	supernatural fact
3rd	I intend that you believe that god is willing [. . . to intervene]	personal religion
4th	I intend that you believe that we want god to be willing [. . . to intervene]	social religion
5th	I intend that you believe that god understands that we want him to be willing [. . . to intervene]	communal religion

Table 8.1

Implications of mentalizing competences (indexed by level of intentionality) for the complexity of religious belief.

After Dunbar (2008).

years ago, belief in a spirit world to which people go when they die, and which living people might visit during trance states, was well developed. Since this identifies only the latest possible date, it is quite likely that such beliefs had been around for some considerable time. However, one thing seems fairly clear: these beliefs, and their associated activities, were found only among AMH populations. There is no convincing evidence for activities of this level of sophistication among the *heidelbergs* or the Neanderthals. This would seem to reinforce the suggestion that religion in this active form evolved only in the modern human lineage and, as such, may well have been a crucial development that allowed them to bind their larger communities.

This conclusion is also implied by the pattern of evolving mentalizing competences shown in Fig. 7.4. The sophistication of one's religion ultimately depends on the level of intentionality one is capable of (Table 8.1). While one can certainly have religion of some kind with third or fourth order intentionality, there seems to be a real phase shift in the quality of religion that can be maintained once one achieves fifth order intentionality. Given that archaic humans, including Neanderthals, don't appear to have been more than fourth order intentional (Fig. 7.4), it seems unlikely that they would have had religions of very great complexity. Quite what that means remains to be determined, but the limited archaeological evidence for an active religious life among archaics suggests that, at best, it wasn't very sophisticated.

The other possible source of evidence for religious belief is the paintings that have been found in some prehistoric

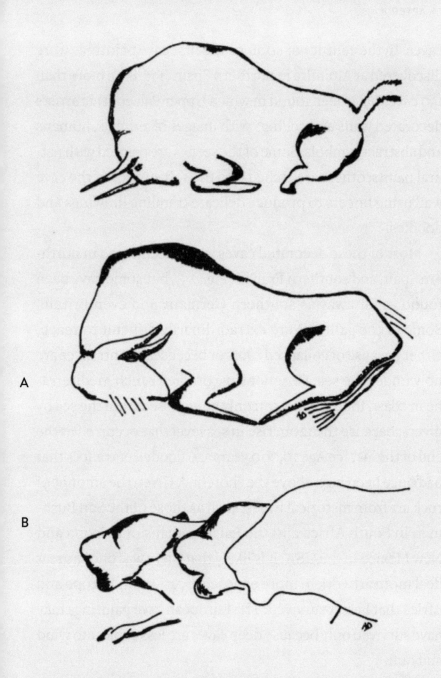

A

B

Figure 8.4

Upper Palaeolithic cave art from Europe. (a) Bison from Altamira cave, northern Spain (dated to 12,000 years ago). (b) The three lionesses from Chauvet cave, southeastern France (dated to 30,000 years ago). Redrawn from Lewis-Williams (2002). © 2014 Arran Dunbar.

caves. In the century or so since the first cave paintings were discovered at Altamira in northern Spain (Fig. 8.4), more than 150 caves have been found in which Upper Palaeolithic artists decorated walls and ceilings with images of animals, humans and abstract symbols. Some of the scenes are painted with natural paints; others are etched into the soft surface of the cave wall using fingers to produce delicate tramline drawings and doodles.

Most of these decorated caves are concentrated in northern Spain and southern France (Fig. 8.5), but some have been found as far away as southern Germany and even Britain. Some of the galleries are extraordinarily difficult to reach, either because of collapsed roofs or because the entrances are now under the sea. Cosquer cave on the French Mediterranean coast, for example, can only be accessed from the sea by divers, because the 120m rise in sea level that occurred at the end of the last Ice Age 10,000 years ago flooded entrances that had once been high above the shoreline. Given the amount of rock art from historical times, such as those of the San Bushmen in South Africa and the Palaeoindians of Arizona and New Mexico in the USA, it is likely that there was once a great deal more artwork in more exposed locations in Europe and Africa that hasn't survived. The European cave paintings may have survived only because deep caves are less subject to wind and rain.

Dating the paintings was difficult until recent improvements in technology allowed tiny samples of the paint to be used. Although some of the paintings are relatively recent (the famous French cave at Lascaux dates to around 15,000 years ago, while the Altamira and Volpe caves in Spain are a

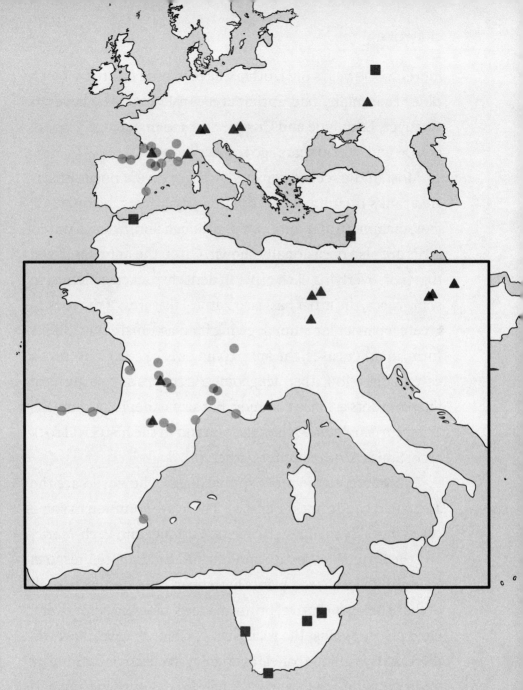

Figure 8.5

Principal archaeological sites that have yielded abstract and figurative art.

- decorated caves;
- ■ personal ornaments;
- ▲ Venus figurines.

After Klein (2000), Bailey and Geary (2009) and Osaka City University (2011).

mere 12,000 years old), others have turned out to be much older than many had anticipated. In France, the caves at Chauvet, Lespugue and Cosquer turned in dates of 30,000, 25,000 and 27,000 years ago respectively.

Most of the scenes depicted in these caves consist of serried ranks of animals, predominantly horses, bison, deer, ibex, mammoth and rhinoceros, although lion, fish and water-birds are also occasionally shown. Often the animals crowd together, overlying each other in densely packed galleries. In some cases, the artist has used natural features of the rock to create a particular animal, giving it an astonishingly lifelike form; in other cases, the animal can only be seen if lit from a certain direction. Abstract geometric shapes and patterns of lines and dots are also common, but scenes depicting humans are surprisingly rare (though common in the historical rock art of South African hunter-gatherers).

Undoubtedly the most spine-tingling, however, are the stencilled handprints (Fig. 8.6). These are common in some caves (notably the French caves at Cosquer and Pech-Merle, and the Spanish caves at San Vincente and Puente Viesgo in Cantabria). In all, 507 prints have been recorded, almost all of these being 'negative' prints made by blowing paint over a hand placed against the rock face. At Chauvet cave, however, there are two unique panels containing, respectively, 48 and 92 'positive' palm prints where the hand was dipped in paint and then pressed against the rock.

Quite why the Upper Palaeolithic peoples did all this artwork is unclear. The large number of animals, some of whom seem to have arrows projecting from their flanks, inevitably led to the suggestion that the paintings depicted

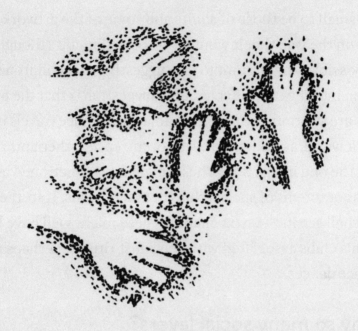

Figure 8.6

Part of a panel of stencilled handprints from Fuente del Salin cave, northern Spain, dated to 20,000 years ago. The images were probably created by blowing mouthfuls of paint over a hand placed on the wall.

Redrawn from Lewis-Williams (2002). © 2014 Arran Dunbar.

magic rituals to ensure hunting success. Many paintings do, indeed, depict animals that were commonly hunted – but by no means all. Some paintings are of predators, like the dramatic lionesses at Chauvet (Fig. 8.4). An alternative suggestion is that the paintings were associated with puberty and other rituals – at least some of the silhouetted hands are far too small to be those of adults and some of the artwork is so low on the wall that it would have been difficult (though not impossible) for an adult to do, suggesting that children were taken into the caves. Yet another suggestion is that the paintings are an attempt to depict shamanistic trance travels in the spirit world, as evidenced by the occurrence of therianthropes like the one in Fig. 8.3a. These might have functioned as the icons or totems of special clubs or associations. If so, then, as we shall see in the next chapter, these might well have been men's clubs associated with specialist rituals, perhaps even trance dances.

Why so many social layers?

I have argued that religion and story-telling became important because of the need to bond communities that were, by primate standards, very large. So let's explore just what these entail in contemporary human societies in a little more detail. As I pointed out in Chapter 3, our natural communities of 150 individuals are not socially homogeneous but rather consist of a series of layers made up of relationships of different emotional and social quality. We have much stronger relationships with some individuals in our 150 than with others, and we see them more often as a result. We have identified four such layers that consist, inclusively, of 5, 15, 50 and 150 individuals, but these layers also extend out beyond the 150 for at least two more layers (500 and 1,500) (Fig. 3.4). Social systems with hierarchically structured layers of this kind are, in fact, characteristic of primates, and hence in themselves are nothing especially novel in humans. What *is* novel in humans is the number of layers. What functions do these various layers subserve? The answer to this question will provide a crucial framework for the fifth transition in the following chapter.

The substructuring of primate groups owes its origins to the need to create coalitions or alliances that protect individuals against the costs of living in large groups. In effect, each layer provides the framework that supports the layer above, with the next layer being an emergent property of the layer below with the two held together in a complex tension. The substructuring of human communities will have arisen for the same reasons: at each level, the smaller grouping makes the existence of the next layer possible.

In some ways, the easiest to explain is the 15 layer, as this is specifically predicted as the natural coalition size based on a relationship between grooming clique size and neocortex volume in primates. In monkeys and apes, the grooming clique (i.e. one's set of regular grooming partners) functions as a coalition to buffer the individual against the stresses of living in large social groups (Chapter 2). As a result, clique size increases in a regular way with total group size. In effect, grooming cliques exist to keep everyone else off your back so as to defuse the stresses of living in groups: the bigger the group, the worse the stresses, and the bigger the grooming-clique-cum-coalition that you need to keep everyone else at bay. The size predicted for humans by the ape equation for this relationship is exactly the 15 layer, suggesting that this may be its principal function in humans. In effect, it is the basis for obtaining social support as well as economic and other forms of everyday social help; it is the set of individuals who will unhesitatingly come to your help when you need it.

The innermost 5 layer that exists within the 15 layer probably functions mainly to offer more intense emotional support. This layer may reflect the fact that humans' more sophisticated mentalizing competences leave them vulnerable to forms of psychological fragility not found in our primate cousins. Theory of mind and higher order mentalizing allow us to imagine the future consequences of our behaviour and hence foresee the terrible things that might befall us in ways that other animals simply cannot do. Having a shoulder to cry on in these circumstances may be crucial to our mental wellbeing and hence our ability to cope with a social world that is much more complex than that in which any other primate lives.

The most obvious characteristic of the 50 layer in hunter-gatherers is that it is the group in which humans prefer to spend the night together, and this may be the key to its original function. Partly because of our poor night-time vision and partly because we have to sleep on the ground, this is the time at which humans are most vulnerable to predators. We are inevitably obliged to rely on group size as the only viable defence against nocturnal predators. While the smaller arboreal primates need to watch out only for day-hunting eagles (mainly hawk eagles or, in South America, harpy eagles), the larger, more terrestrial primates are more susceptible to nocturnal predators: the three most important predators of baboons are leopards (nocturnal), lions (hunt mostly at night) and hyenas (nocturnal). While it is true that the 50 layer also provides the basis for women's foraging (women tend to forage in rather larger groups than the men), it seems unlikely that humans needed groups of this size just for digging up roots or collecting berries. On balance, their role seems to be protection against predators at night, and perhaps to provide a pool of people for safety while foraging during the day.

We have identified six plausible functions for the 50 and 150 layers: protection against predators, defence of territory or food resources, defence of reproductive mates, trading arrangements to minimize environmental risk, information exchange on the location of resources, and protection against raiding by neighbouring human communities (otherwise known as the warfare hypothesis). In a series of studies, we came to the conclusion that what has become known as the warfare hypothesis (defence against raiders) had by far the strongest support as an explanation (Table 8.2). It may be no accident that the social brain equation predicts that the 150 layer appears for the first time in the archaeological record (Fig. 3.3) at more or less the time that we see the beginnings of the demographic explosion around 100,000 years ago.

Of course, once you have a bonded community to act as protection, you can also use it for multiple other purposes. Trading arrangements that allow you to seek refuge elsewhere when things are bad in your local area are an obvious

ORGANIZATION LEVEL	SIZE	PREDATION PROTECTION	RESOURCE DEFENCE
Family	~5	no	no
Lineage	~15	(maybe)*	no
Band	~50	yes	no
Community	~150	no	no
Mega-band	~500	no	no
Ethnolinguistic (tribe)	~1500	no	no

(maybe)* indicates a function that is likely to be an emergent property (i.e. one that becomes possible once the layer is in place, but is unlikely to have been a factor selecting for the layer).

RESOURCE TRADING	MATE DEFENCE	INFORMATION EXCHANGE	RAIDING CONSPECIFICS
no	no	no	no
no	no	no	no
no	no	(maybe)	yes
yes	no	(maybe)	yes
yes	no	yes	yes
yes	no	yes	yes

Table 8.2

An assessment of the most likely functions for each of the six grouping levels in modern hunter-gatherer societies.

Adapted from Lehmann et al. (2014).

secondary benefit for large communities, and the same is surely true of the benefits that a large, more widely distributed community provides in terms of information about the distribution of food resources. Exchange networks have been described in many hunter-gatherer societies. Among the Ju/'hoansi of southwest Africa, for example, the exchange of 'symbolic' gifts known as *hxaro* creates a dispersed network of mutually supportive relationships. *Hxaro* partners provide help, often by allowing exchange partners to seek refuge with them when conditions deteriorate in their local area. The anthropologist Polly Wiessner has described cases in which half of one San population moved in with distant *hxaro* partners during a period of food shortage that would have resulted in certain death had they not done so. Similar exchange networks have been reported for other hunter-gatherers, and there is some suggestion that they might have existed in the European Upper Palaeolithic.

The important feature of *hxaro* partners is that they are not members of your own (50-level) band (the people you actually live with), but people who live elsewhere. In Wiessner's San population, *hxaro* partners typically lived within 40 km of each other, which, as the anthropologists Bob Layton and Sean O'Hara have shown, just happens to be the average radius of a community territory in tropical hunter-gatherers. In other words, *hxaro* exchanges take place principally between people who belong to the same (150-level) community, reinforcing the suggestion that the community is the core unit for the exchange cooperation, not the band (as some researchers have mistakenly assumed: see also Chapter 3).

We should also consider the layers beyond 150 that we know, from our analyses of ethnographic societies, form a natural extension of the scaled series of grouping circles in the contemporary world (Figures 3.2 and 3.4). It has been suggested that the 500 layer is the minimum size to maintain genetic exchange without risk of excessive inbreeding: in many traditional societies, people generally avoid marrying members of their own (150 layer) community, and instead typically marry people from the adjacent community (the partners in their 500-layer extended community). It seems a perfectly sensible explanation for this layer, since it consists mostly of people we would consider to be acquaintances – those with whom we have semi-casual relationships that are not based on reciprocated, personal friendships but rather based on more formal, categorical relationships (which, of course, are wholly dependent on language). That is potentially important from a mating point of view because it means that our community knows them well enough for one of our network members to be able to vouch for their suitability as marriage partners.

That leaves only the 1,500 layer, conventionally identified in the ethnographic literature as the tribe or ethnolinguistic community, a community defined by the fact that everyone speaks the same language. The most likely explanation for this layer is that it represents a trading network that allows communities to buffer themselves against the risk of environmental turbulence because it covers an area large enough to access a range of habitats. The sample of tropical hunter-gatherers compiled by Layton and O'Hara had an average

150-community territory size of 5,000 km², so the 9–10 communities that would be included in the tribe would have a combined territory of around 50,000 km² (an area roughly 225 km by 225 km) to draw on in times of ecological collapse within their own territory (or, for that matter, raiding by other groups). That would provide an area large enough to ensure plenty of refuges in times of disaster, since catastrophes are rarely so widespread.

As such, it seems plausible to suggest that tribal level groupings probably arose by extending the subsidiary trading arrangements from 150-level communities to ever wider circles as environmental conditions became progressively more stressful during the last Ice Age beginning around 100,000 years or so ago, and coinciding with the last increase in brain size. It is likely that this layer was crucial in allowing modern humans to colonize high-latitude habitats in Eurasia as successfully as they did after 40,000 years ago. Maintaining such extended social networks probably depends on shared culture and moral views (to ensure honesty and reliable reciprocity), and so may have been unique to AMH. Here, then, may lie the origins of Deacon's symbolic community, with their ability to negotiate contracts. Being able to draw on such an extended network for support may also have been critical whenever modern humans and Neanderthals fell into conflict: Neanderthals may have been simply unable to cope with the overwhelming odds that AMH groups could call up when they needed to.

The developments described in this chapter mark the final flowering of human culture, the high point to which 6 million years of biological evolution has led us. Within this phase, however, lie the seeds of the fifth and last transition, for everything we have discussed in this chapter has been cultural, and culture is learned and the product of curiosity and invention. The fifth and last transition, heralded by the Neolithic, marks the point where cultural invention takes over from biological evolution. It is the subject of the next chapter.

CHAPTER 9

The Fifth Transition

THE NEOLITHIC AND BEYOND

Towards the end of the last Ice Age, around 12,000 years ago, humans in the Levant began to settle in villages, building more permanent houses made of mud brick and other materials. It would be several millennia before they 'invented' agriculture in what is known as the 'Neolithic Revolution'. Indeed, it is likely that the stimulus to start growing plants was the need for more resources in order to be able to feed the increasing numbers of people in a settlement. Why they started to settle in the first place remains a mystery. One obvious possibility is that settlement started in order to provide protection against raiding by neighbours. Even if this wasn't the case, raiding almost certainly became a major issue quite soon, since it is thought to have been responsible for the very rapid rate at which settlement size increased over the next 5,000 years or so, leading eventually to the foundation of city-states and petty kingdoms.

By 8000 BC, settlements like Çatal Hüyük and Göbekli Tepe in Turkey and, later, Jericho in the Levant had become significant walled towns covering as much as 15 ha with 1,000 dwellings and perhaps as many as 5,000 inhabitants. Within a few thousand years, settlements of this kind were widely distributed across southern Eurasia (Fig. 9.1). These structures

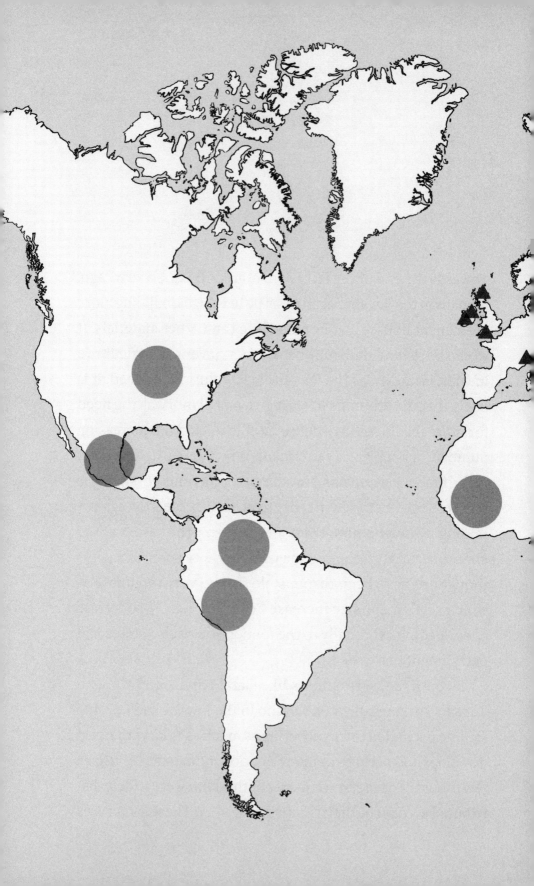

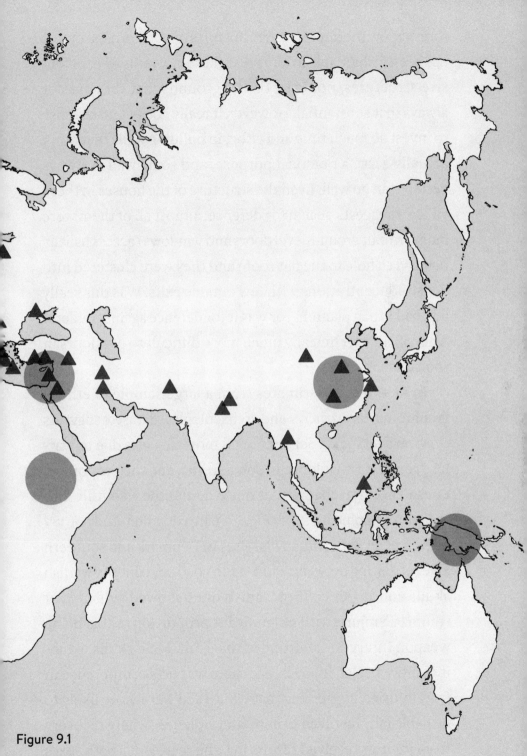

Figure 9.1

Main Neolithic settlement sites (▲). Grey circles indicate main
locations for domestication of plants and animals. After Diamond (2002).

foreshadow the later Iron Age hill forts of northern Europe. It has been fashionable to deny that these city walls were defensive structures, not least on the grounds that they are not always that substantial. However, it really doesn't make sense to invest so much time and effort in building walls that don't actually serve a practical purpose, and some wall is always better than no wall. Even the structure of the houses in these villages suggests design for defence: almost all of them were built without ground-level doors and windows (access usually being via a hole in the flat roof) and they were clustered into dense concentrations with few outside exits. Was this really just bad urban planning, or was it for defence against raiders? After all, it is technically much more difficult to put doors in roofs than in walls.

In fact, recent estimates from a large sample of ethnographic hunter-gatherer and archaeological datasets suggests that, on average, 15 per cent of all mortality was due to warfare (or at least violence in some form), and the proportion doesn't seem to change that much across the Neolithic into the ethnographic present (Fig. 9.2). In a detailed study of 387 burials from 83 Scandinavian sites in Denmark and southern Sweden dating between 3900–1700 BC, 9 per cent of Swedish skulls and 17 per cent of Danish ones showed evidence of traumatic injury, with a significant proportion of this being weapon injury to the front of the skull. Male skulls exhibited more healed injuries, but there was no sex difference in fatal injuries, suggesting that while men were more likely to be habitually involved in non-fatal violence (where recovery from injuries received results in bone regrowth), both sexes were equally likely to die from violence.

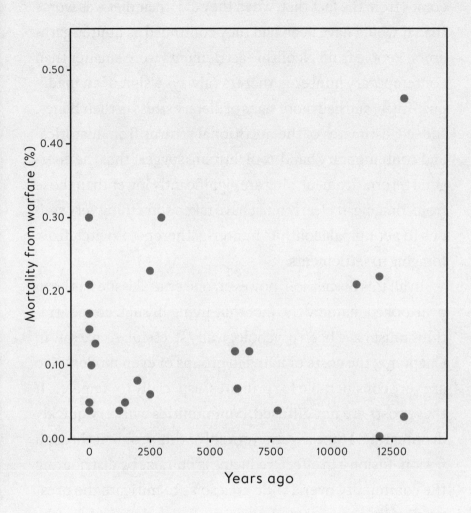

Figure 9.2

Percentage of adult mortality due to warfare in a number of ethnographic and prehistoric societies, plotted against their dates. The data are based on skeletal samples from burials.

Source: Bowles (2009).

Perhaps the most compelling evidence that people did not gather together in these settlements entirely voluntarily comes from the fact that, when they did, their diet was worse than it would have been had they continued as hunter-gatherers. People from Neolithic settlements were smaller than contemporary hunter-gatherers (always a sign of an inadequate diet) and had more signs of dietary stress in their bones. Indeed, estimates of the nutritional returns from historical and contemporary hand-tool farming suggest that the energetic returns from farming are significantly lower than those from foraging. It clearly must have taken something very serious to get Epipalaeolithic[1] hunter-gatherers to switch from foraging to settlements.

In all this discussion, however, one crucial issue has been overlooked, namely the fact that living in such concentrations must have been psychologically stressful. As we saw in Chapter 2, the costs of living in groups of even modest size are very considerable for primates, especially for females. If these costs are not diffused, communities will very quickly disintegrate. Hunter-gatherers solve this problem through fission–fusion – in effect, reducing group size by distributing the community over a wide area so as to mitigate the pressures of living together. But fission–fusion doesn't solve the whole problem. Hunter-gatherer sociality is a compromise between the pressure to live in groups (defence against predators and/or human raiders) and the pressures to disperse. In the last three chapters, I suggested that music (singing and dancing), then language-based story-telling and finally religion evolved as mechanisms to neutralize these stresses and facilitate social bonding in modest-sized communities.

The problem that the Neolithic sets for us is not that humans invented agriculture, discovered how to store foods or learned how to build houses – these are relatively trivial accomplishments of the kind that humans can solve easily – but, rather, how humans managed to solve the genuinely destructive problem created by living in very large, spatially cramped settlements. Solving this problem was crucial to the later development of urbanization and the rise of city-states: without finding a solution, none of these would have happened. The fifth of the great transitions that had to be managed on our way to becoming modern humans was the Neolithic Settlement, with the emphasis on 'settlement', not 'Neolithic'.

Solving the collective action problem

Primate societies, as I explained in Chapter 2, are implicit social contracts intended to solve the problems of survival and successful reproduction communally by sharing the costs involved. By grouping together in stable groups, monkeys are able to deter predators from taking so much interest in them when there are easier prey to be had. Defence doesn't need to be active: it is the deterrence effect of a large group that is of most importance. The difficulty for all such social contracts is that they are susceptible to freeriding, whereby some individuals take the benefit of being part of the contract but don't pay all the costs: they hang back when danger threatens so as not to put themselves at risk, or they don't invest time and effort in whatever activity is necessary to gain the benefit (taking turns to keep watch, grooming others).

Much has been made of social punishment[2] as a mechanism for reducing freeriding. However, punishment is a problem from an evolutionary point of view: the punishers are behaving altruistically, and this puts them at an evolutionary disadvantage in terms of conventional individual-level Darwinian selection relative to those members of the community who hang back and allow others to do the punishing. In effect, the non-punishers are freeriders who exploit the altruism of the punishers, which just moves the whole problem one step further back. Group selection (whereby a trait evolves because it benefits the group at the expense of the individual) is not the answer, because, in all conventional selection environments, the altruists would always lose out and be selected against. Kin selection might work if the group consisted entirely of kin, but, realistically, kin selection will only work if the beneficiaries are close kin because relatedness declines very rapidly once you are past first cousins.

Punishment certainly works, but it is not necessarily the best way of dealing with the problem. When the chances of being caught are small and the community is large (so that individuals don't feel that they are under any obligations towards each other), then punishment often doesn't work, no matter how draconian the penalty may be. In conservation biology, this is known as the 'poacher's dilemma'. The penalty for being caught can be very substantial, but if the risk of being caught is low, then it pays to poach (or over-fish the high seas, or cut down the tropical forests for timber), even when we know we would all be better off in the long run if we all stuck to the common bargain. Humans are rather bad at what economists call 'future discounting': we tend to prefer an immediate reward rather than one in the future, even when the future reward is greater – in other words, the ability to inhibit impulsive (or prepotent) responses, which, as we noted in Chapter 3, is correlated in primates with the size of the prefrontal cortex.

A psychologically more effective way of ensuring that people stick to the rules of the community is if they are themselves committed to it and so have a sense of obligation to the other members of the group. Kinship is obviously one mechanism that creates this sense of commitment. An alternative way of engineering this is to make people pay a cost for belonging to the community. Fig. 9.3 illustrates this for nineteenth-century American utopian communes: the length of time for which a commune survived depended on how demanding it was in what it asked you to give up in order to be a member. It seems that the more you have to give up, the more willing you are to put up with the petty squabbles and fractiousness of other members, and so the longer your commune will last. But notice that this was only true of religious communes. Not only did secular communes not last as long as religious ones, but they also did not show this commitment effect. It seems that there is something especially engaging about having a spiritual dimension to the community ethos that makes us more willing to hold back on our more disruptive and selfish behaviours and adhere to the community line.

We use a number of mechanisms to create a sense of community. In many ways, the most important of these is a shared worldview. If I see the world as you do, speak the same language, have the same moral attitudes and believe in

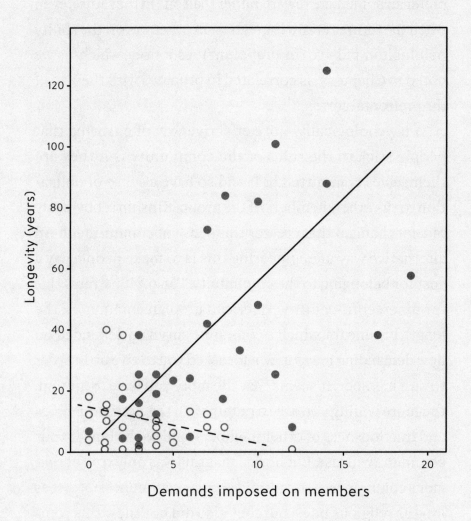

Figure 9.3

Length of time for which nineteenth-century US utopian communes survived after they were founded, plotted against how demanding they were of members (indexed as the number of things members had to forgo or do in order to belong). Communes with a religious basis (●, solid line) typically lasted longer than communes with an exclusively secular philosophy (○, broken line), and their survival times increased as a function of how demanding they were. Redrawn from Sosis and Alcorta (2003).

the same explanations for why the world is as it is, it signals that we probably grew up in the same small community. And that makes it more likely that we can trust each other. And in small communities, one reason we can trust each other is that we are very likely kin. Other cultural markers such as common dress codes, hair styles, pottery designs and dialects also signal membership of a community.

There is, however, another aspect of human behaviour whose importance in this respect has been almost completely overlooked, and this is feasting. There is now growing evidence for feasting in the archaeological record from the Natufian period (13,000–9,800 years ago) and the PPNA (Pre-Pottery Neolithic A: 8,000–7,000 years ago) in the Levant. The evidence includes both the remains of many very large animals (aurochs, or wild cattle) that were far too large to have been mere family dinners and the presence (e.g. at Göbekli Tepe in southeastern Turkey) of very large stone vats (some capable of holding 160 litres, and far too heavy to move easily) that can only have been used for brewing. Indeed, there is direct evidence of brewing from the residues in some of these vessels. Alcohol thus seems to have had a very ancient history in human affairs. As a matter of fact, the kind of primitive barley that was first cultivated was not especially suitable for bread (it yields a small harvest and the glume adhering to the grain reduces its value), but was quite nutritious when turned into a gruel or mash – not the most exciting thing to eat, but perfect for beer-making. Barley and other einkorn (primitive wheat) may thus originally have been cultivated to produce beer and not bread.

As it happens, alcohol triggers a massive endorphin

surge. In fact, it is not the alcohol that alcoholics become addicted to, but the endorphins, which is why naltrexone (a β-endorphin antagonist)[3] is used in the treatment of alcoholism. This may explain why drinking alcohol together is universally considered a social bonding activity. I remarked in Chapter 6 that eating also triggers a strong endorphin effect, and this may explain why social meals and feasts seem to play such an important role in our social life. It may well be that they appeared during the Neolithic as a way of both bonding the community and also welcoming visitors (and especially strangers) into the community. Inviting someone for dinner (with or without alcohol) remains an important feature of modern social life – yet no one ever comments on how odd this is as a behaviour or asks why it evolved. The answer seems to be that feasting provides an important endorphin punch that helps develop social bonding.

Stable family, fragile friends

Kinship undoubtedly smooths the processes of social cohesion significantly because family are more tolerant of each other and more willing to help each other when the need arises (the so-called 'kinship premium' that arises from the evolutionary process of kin selection). However, increasing community size beyond 150 inevitably increases the number of unrelated individuals whom one is likely to meet regularly because, as I pointed out in the last chapter, the 150 layer more or less demarcates the meaningful limits of kinship. Once settlements grew to include the layers beyond this, and eventually even complete strangers, something extra was needed to ensure that the community as a whole was sufficiently bonded.

We may be able to gain some insight into how we created the larger communities of the Neolithic by examining how we create friendships with unrelated individuals today. We service friendships by interacting socially. Traditionally, this would mean face-to-face contacts. In a study of how relationships change over time when the opportunity to meet is reduced by physical distance, Sam Roberts and I asked participants to rate their emotional closeness to each of their family members and friends over an 18-month period after moving away from home. Emotional closeness to kin remained remarkably stable across time (and may even have increased a little), but emotional closeness to the original set of friends from the start of the study dropped off precipitously when they no longer had the opportunity to meet so easily (Fig. 9.4). In other words, friendships with unrelated individuals have high maintenance costs and decline in quality rapidly when not serviced. This has important implications because in order to maintain coherence in a larger community that extends beyond the 150 layer, it will be necessary to include many high maintenance non-kin relationships and this will inevitably put pressure on time budgets.

In contemporary personal social networks, the 50 layer is disproportionately populated with friends whereas the outermost 150 layer is disproportionately populated with extended family. Because extended family require less investment – less social capital – than friends, the outermost layer is much less demanding to maintain than the inner layer. If we

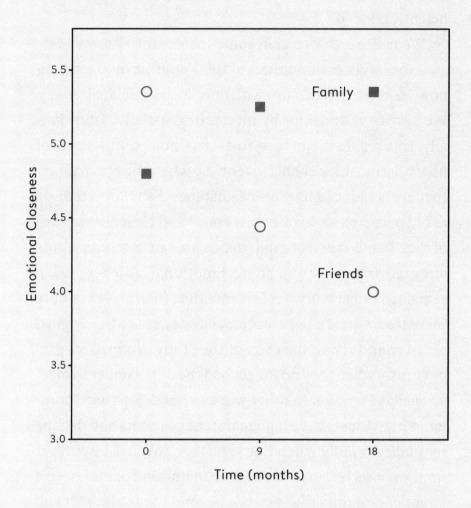

Figure 9.4

Subjects rated their emotional closeness (on a 1–10 scale) to all members of their social networks at three points over an 18-month period. At month 6, most subjects moved away from home and were unable to contact the members of their original social network as easily as they had done previously. The plotted data are the average values to all extended family members and to the original set of friends that they had at the start of the study (month 0).

invested as little time in friendships as we invested in family members in this outer layer, the friends would be little better than acquaintances – people we are on nodding terms with, who are much less likely to come to our help when we need it. Reversing the ratio of friends and family between these outer layers would significantly increase the costs of maintenance and will inevitably reduce the level of coherence and commitment in the network. In fact, this may be why friends are concentrated in the 50 layer: the frequencies of interaction with people in the outer layers is just too low to maintain functional friendships, and friends who slip into this layer quickly drop down through it into the acquaintances (500) layer beyond.

While we are stuck with relatives whether we like them or not, we do at least choose whom to make friends with. So whom do we choose as friends? We asked a large number of people to think of a particular friend in each of the three outer layers of their social network (the 15, 50 and 150 layers), and then say whether or not they shared any of a long list of traits with each of them. The results suggest that there are six main dimensions to friendship: shared language, shared place of origin, similar educational history, shared interests and hobbies, shared worldview (similar political views, religion and/or moral attitudes) and shared sense of humour. Sharing any two or more of these creates a friendship of a given level of emotional closeness; sharing more creates a proportionately stronger one. These six dimensions seem to be pretty much substitutable: any three of them will yield a three-star friendship, and it doesn't matter too much which three dimensions you happen to share. More importantly, your willingness to behave altruistically towards a friend, to help them out in time of need, also depends on the number of these trait-dimensions you share. Thus, it seems that, despite what we suppose, friendships are born and not made. We may have to make do with the ones we find, but the quality of the friendship and how long it lasts will be the poorer the fewer traits we share.

Notice that all the traits that underpin friendship are cultural in origin: none of them is biological and fixed. This is important precisely because cultural traits identify the community we come from: sharing cultural traits of this kind provides cues of which community we were born into. More importantly, it allows these badges for community membership to change over time, which itself helps to signal that your membership is up to date. It also helps, as Daniel Nettle and I showed in a computer model, to control free-riders. If you know the style and can walk the walk, it must be because you grew up there; this makes it difficult to come into a community as a stranger and learn the local 'style'. Signalling community membership provides a guarantee of your honesty because the community itself stands guarantor for your behaviour.

In a simple test of this, Oliver Curry and I asked subjects to rate the degree of interconnectedness between their eight closest friends (in effect, their 15 layer) and then say how likely they would be to act altruistically towards each of them. Those whose friendship networks were densely interconnected (they saw more of each other and went out together as a group) were more altruistic towards their friends than those whose networks were less densely connected. At one level, this is just a reputational issue. The network watches, and monitors its members' behaviour: failure to offer help or repay past favours are noted and commented on. We act altruistically towards our friends for fear of the reputations we may gain and the fact that they will shun those whose reputation becomes tarnished. But there may also be a positive aspect to this: when you see a lot of each other, and do a lot of bonding activities together, your relationships will be stronger and the positive feelings this generates may make you act more prosocially towards each other. You may not even need the threat of punishment.

In the course of our studies of friendship, we noticed important differences between the two sexes in the way their social worlds seem to work. Two are especially relevant. One is the fact that women often have an additional very close friend (usually, but not always, a girlfriend, now sometimes referred to as the BFF, or 'Best Friend Forever') as well as a (typically male) romantic partner, whereas men don't. Men's friendships seem much more casual, and so tend to involve more friends at the same looser level. This has a number of important consequences in terms of relationship stability. Close female relationships play an important emotional support role for women, but they are also more fragile: when they break down, they often do so catastrophically and are beyond repair, whereas men's more casual relationships are usually repaired more easily. Women make great efforts to maintain their close friendships, whereas men often simply walk away from them – out of sight means literally out of mind, such that, after moving away, they simply substitute a new set of friends wherever they happen to land up.

In effect, women typically have a few rather intense friends, whereas men tend to have many rather casual friends – or, to put that another way, men find clubs rather congenial. This tendency for men to like clubs with all their paraphernalia and bonding rituals (something that women often find puzzling) seems to point towards the fact that men rather easily form groups or gangs. Although this may have its origins in small hunting (or, more likely perhaps, war) parties, the psychological mechanism involved seems to be of a kind that can easily be generalized to much larger numbers and so lends itself to the formation of large hierarchical organizations like churches or armies.

The second difference between the sexes is in the way relationships are maintained. In our 18-month study of friendships, we investigated the mechanisms that prevent friendships declining over time. We asked participants to say how frequently they contacted each of their friends (either face-to-face or by phone or email) and how often they had done things together (gone shopping, been on holiday, gone clubbing, helped with house moves, etc.). What prevented women's friendships deteriorating was talking together, whereas what prevented men's friendships deteriorating was doing things together. Talking together had absolutely no effect on the survival of men's friendships.

The important conclusion from this brief excursion into the nature of friendships is that the task of maintaining social cohesion in super-large communities that exceed the conventional hunter-gatherer community of 150 was not a trivial problem. Something else was needed, and the seeds for this appear to have been sown in the way human friendships are naturally organized: the creation of clubs, and in particular men's clubs. Clubs based on a single theme may be all that is needed. The origins of these single-interest clubs probably lie in kinship, since the outer layers of kinship networks are purely language-based and one-dimensional. Religion probably provides a second early example that builds on this foundation.

Networks on the edge of civilization

The archaeologist Fiona Coward used the similarities between the artefacts (pots, grindstones, statuettes, jewellery, etc.) of different communities in the Levant during and just before the Neolithic to gain insights into networks within and between these early communities. She found that, as community size increased, so their connectedness declined: their artefacts began to look less and less similar. Nonetheless, there was no evidence of internal fragmentation. The reason why fragmentation didn't occur might lie in an increasing complexity of the material culture (new kinds of artistic representation or personal ornamentation) over time, which may have offset the natural tendency to fragment by providing more cultural complexity to define the community. In other words, the rising complexity of the material culture may have been a deliberate response to the risk of fragmentation as community size increased rather than being an accidental by-product of community size. In this respect, it seems to parallel the Upper Palaeolithic Revolution explosion in Europe some 20,000 years earlier. Indeed, in the Levant, many of these changes seem to have kicked in during the middle and late PPNB (Pre-Pottery Neolithic B, dating to around 8,000–6,000 years ago), just as the Neolithic was gathering strength.

The Neolithic offers one important change in respect of religion that might just provide the key to the transition. Historians of religion have long distinguished between two radically different kinds of religion among the many thousands that one finds around the world today: shamanic and doctrinal religions. The first, which we considered in some detail in Chapter 8, are religions of experience; the second are religions associated with sacred spaces (e.g. temples or churches), priestly hierarchies, theologies, gods (sometimes, but not always, a 'High God' who regulates the lives of men) and formal rituals that are used to appease the gods in order to bring good fortune. This second type of religion is very different from shamanic religions both in style and in performance. The formation of permanent settlements seems to be the transition point between these two forms of religion: even today, nomadic and semi-nomadic hunter-gatherer and herder societies the world over are characterized by shamanic religions, whereas societies that live in permanent settlements are invariably characterized by doctrinal forms.

It would be easy to interpret the transition between these two types of religion as being a consequence of the fact that it is simply so much easier to have ritual buildings once you live in permanent villages. However, there is no obvious reason why you need to have temples just because you live in a village. Living in villages is no real impediment to having a shamanic-type religion: indeed, some settled cultivators like the Hopi of the southwestern US still have shamanic religions. Since shamanic religions typically involve music-based dancing, all one really needs is an empty space (a dance floor or parade ground) where one can dance. Conversely, there is no real reason why nomadic hunter-gatherers can't agree on a central sacred place. Many hunter-gatherers, like the Australian Aboriginals, have sacred places (Uluru/Ayers Rock being a case in point). There is even archaeological evidence that temple-like structures at Neolithic sites like Göbekli Tepe were first built before the local population built permanent houses. It is possible that people lived around these ritual sites in temporary dwellings that simply haven't survived. How long such a phase lasted would depend crucially on the amount of pressure such communities faced from raiders. The problem is that people gathered together in one place are more at risk of attack because they are easier to locate, and this would likely make such a phase rather short-lived.

The creation of special buildings used solely for ritual purposes implies a deliberate shift to some kind of collective ritual that is very different to the less formal practices that characterize shamanic religions. Formal ritual buildings imply priestly specialists who intercede on the community's behalf and who enact special rituals from which ordinary individuals are excluded as mere onlookers. Half a century ago, Raoul Naroll demonstrated, from an analysis of data for small-scale societies, that specialist roles (potters, craftsmen, victuallers, mercenaries, as well as priests and administrators) start to emerge once community size exceeds about 500 individuals. This may identify a natural crisis point at which a phase change in organization has to occur if community coherence is to be maintained.

At least in terms of modern religions, the shift from shamanic to doctrinal religions marks another important change: a switch in both the intensity and frequency of religious events. In shamanic religions, trance dances or their equivalent occur at irregular intervals (typically, when someone thinks one is needed) with a cycle time that commonly approximates a month or so; in doctrinal religions, religious services are less emotionally intense but occur at more frequent intervals (commonly a weekly schedule). This suggests that, while trance states are very effective at diffusing tensions, the intensity of the experience is such that they would be stressful if repeated too frequently. But if living in larger communities radically increases the stresses, then a shorter interval between religious events may be needed – and a shorter interval may only be possible if the religious events are less intense. One obvious way of making religious events less intense is to separate specialists who perform the rituals (individuals trained to take the pressure, as it were) from the congregation who merely observe.

The fact that most of those taking part do not experience the magnitude of uplift provided by trance states means that something else may be necessary to enhance the community-bonding effects of religion. It is surely no coincidence that High Gods who take a direct interest in the activities of humans, demand certain standards of behaviour and have omniscient oversight of human behaviour start to play a significant role only in doctrinal religions.[4] Indeed, in contemporary tribal societies, the presence of a High God (or gods) is correlated with the size of the community (the bigger the community, the more likely it is that it will have a religion with a High God) and a number of studies have shown that, within these kinds of societies, people who actively believe in the God are more likely to act prosocially and adhere to communal rules than those who do not. Even in the contemporary USA, states that have higher church attendance (i.e. whose members are more actively religious) have higher levels of social engagement and less crime than states whose citizens are less religious, as Robert Putnam eloquently showed in his seminal book *Bowling Alone*.

At the same time, we should not underestimate the extent to which religious beliefs, often encompassing a complex mixture of a worldview, origin stories and moral codes, create a sense of belonging and community membership. In this respect, this seems to build directly on the basic dimensions that characterize friendships. It is as though the fundamental psychological processes that underpin friendships have been exploited during the historical evolution of doctrinal religions as a basis for creating a sense of membership of a virtual community that includes very large numbers of complete strangers. The importance of kinship may be noted here too: almost all doctrinal religions exploit the language of close kinship (terms like *father*, *mother*, *brother*, *sister*), apparently in an attempt to create an illusion of family relatedness.

In my view, the Neolithic Revolution was, as much as anything, a religious revolution. It is marked by a switch from more casual forms of shamanic religion to more organized forms of doctrinal religion in which discipline is imposed on the members from above by a hierarchy of religious specialists who claim to be acting on behalf of the spirit world – which itself now takes on a form that is more defined, involves specific individuals (gods, saints) and is universally accepted by all adherents.

This switch to doctrinal religions has, however, been less than perfect. All the doctrinal religions are beset with problems of maintaining coherence in a way that perhaps reflects their origins. Despite their hierarchical structure and the threat of punishment in this world or the next, all doctrinal religions experience a veritable torrent of breakaway cults and sects. This is rather similar to the way dialects form in languages, which, as we saw in Chapter 8, are also designed to reinforce small-scale community allegiances by differentiating one community clearly from the next. It seems that the doctrinal religions have never quite managed to break free of their origin in shamanic religions. Because these cults are invariably more mystical and ecstatic, based on emotional engagement rather than intellectual commitment, they threaten to undermine the theological and political control that the priesthood exerts; as a result, most of the doctrinal religions (Hinduism is probably the only real exception) have historically taken active steps to suppress what they regard as heretical sects.

Some of these breakaway cults eventually become world religions in their own right – Christianity and Islam out of Judaism, Orthodoxy and the Protestant churches out of Rome, the various Shia, Sunni and Sufi branches out of Islam, the different schools and sects of Buddhism, to name but the most familiar. It seems that the doctrinal religions can never quite break free of these ancient forces: small, personal, experiential, shamanic style religious groups constantly bubble up from deep beneath the surface. What seems to be central to cults is the emergence of a charismatic leader (typically, but by no means always, male) who becomes the symbolic and functional leader around which a small community of adherents gathers (often, but not always, women). As such, the personality and style of the individual leader play a much more important role in the initial success of cults than the particular theology they advocate. Familiar recent examples range from the Maharishi to David Koresh to the Reverend Jim Jones (he of the Jonestown massacre fame).

Deacon's dilemma

There is one final issue we need to address. The increasing size of settlements through the Neolithic exacerbated one aspect of human social behaviour that has major implications for social cohesion. At some point during the course of our evolution, we developed a form of romantic pair bonding that continues to play an important role in both reproduction and social arrangements. As Terry Deacon observed in his book *The Symbolic Species*, this creates a potentially explosive problem once there is some form of division of labour: if males are often away from camp, they will be unable to guard their spouses from rivals who might sneak matings with them. Such illicit matings can be beneficial for the females (they may gain better quality genes for their offspring or access to additional resources), but disadvantageous for the males who risk devoting time and resources to rearing another male's offspring – a form of altruism that is evolutionarily detrimental. I refer to this as 'Deacon's dilemma'. It is a problem whose intrusiveness and destructiveness will be infinitely worse in the large, more anonymous communities of Neolithic settlements.

Deacon's solution, as we saw in Chapter 8, was to suggest that humans evolved formal language-based marriage contracts. That way you can signal who is 'spoken for' and so should be left alone. Whether marriage contracts and public signals like wedding rings have ever prevented adultery is debatable (the answer is: probably not), but the fact is that humans do commonly signal marriages in this way and it may have some deterrent effect even if it is not perfect. More importantly for our present concerns, however, is Deacon's central point: sexual competition and jealousy can be a very destructive force not just at the level of the pair but more generally for the community as a whole when a couple fall out or a third party hoves into view, especially in small-scale societies. While it is possible to manage and ameliorate the level of jealousy to some extent through cultural means, these never work completely: the green mist invariably seeps through the interstices of the cultural mesh, followed very closely by the red mist. It may not have been mere prudishness that led to many of the nineteenth-century American utopian cults banning sex between their members (the Shakers were a classic example) but rather an intuitive understanding of its potentially disruptive effects.

This inevitably raises the hoary old question as to whether humans are really monogamous or polygamous. I don't want to rehearse all the arcane arguments on this topic, because most of these miss the point completely. Humans have *pair-bonded* mating relationships irrespective of whether they happen to opt for monogamous or polygamous (or for that matter polyandrous) marriage arrangements. In other words, the romantic relationship itself doesn't necessarily depend on the marriage system.[5] Marriage systems seem to be quite flexible and depend mainly on the local economy and cultural traditions.

That said, on all anatomical indices that distinguish obligately monogamous primates from polygamous ones, humans invariably sit – somewhat inconveniently for both sides of the argument – exactly on the cusp between the two. In Chapter 4, we saw that ape species that mate polygamously or promiscuously have low 2D:4D digit ratios (reflecting high levels of circulating testosterone during gestation), whereas obligate monogamists like the gibbons have ratios that approximate equality, with modern humans falling exactly half way between the two (see Fig. 4.5). The same is true of sexual dimorphism in body weight: we have a modest degree of sexual dimorphism (men are about 8 per cent taller than women and 20 per cent heavier), and fall roughly midway between the monomorphic monogamous species like gibbons and the highly dimorphic polygamous species like baboons and great apes (where males weigh between 50 and 100 per cent more than females). The relative size of male testes also provides an anatomical index of the mating system in primates: the males of polygynous species like chimpanzees have very large testes for their body size, whereas those of monogamous species like gibbons and strict harem-formers like the gorilla have very small ones.[6] Once again, modern humans sit ambiguously in the middle of the two distributions, indicating a mild degree of polygyny. In short, what we need to explain in modern humans is not monogamy but romantic pairbonds – which, contrary to the protestations of social anthropologists, are in fact universal across human cultures.[7]

So why should pairbonds have evolved? In Chapter 2, we saw that three explanations have been offered for the evolution of monogamy (i.e. pairbonding) in primates, of which only one (infanticide risk) received any real support. Females opt for monogamy as a mechanism to reduce the risk of harassment by males in what has become known as the 'hired gun (or bodyguard) hypothesis'. In some cases, such as the gorilla, several females attach themselves to the same male, so creating a polygamous mating system. But in most cases, each female attaches herself to a different male to form a genuinely monogamous pair.

In considering hominins, we can simplify the issue to a contrast between just two hypotheses: biparental care and infanticide risk. The third hypothesis (that males are forced to be monogamous because females are so widely spaced that males cannot defend mating access to more than one at a time) simply doesn't apply to humans, because human females do not range alone in individual territories, and, as we have seen, females almost certainly never have done in the history of hominin evolution.[8] Human females are firmly social and always live in groups. Biologists, anthropologists and archaeologists alike have invariably assumed that pairbonds evolved in modern humans to make biparental care possible, mainly on the grounds that in most modern societies fathers provide some paternal care – or, perhaps more convincingly, provide food for their spouse and children by hunting. But there is a major difference between the factors that select for something and those that kick in afterwards as a result of a window of opportunity that arises *once* a trait is in place. In Chapter 2, we saw that, in primates, biparental care (or more explicitly, paternal care) invariably evolved *after* pairbonding, and there are no principled reasons for supposing that the situation has ever been any different in humans.

More importantly, as the anthropologist Kristen Hawkes has pointed out, it is far from obvious that hunter-gatherer males actually do contribute significantly to parental care by providing food for their families. Sometimes they do, and sometimes they don't, and at least some of the activities that look like parental care (namely, hunting large game) are better explained as advertising for mates.[9] Although her suggestion has been much argued about, the evidence for male parental care in humans is not exactly overwhelming and needs a lot of massaging to make it look even plausible, never mind convincing. If males can get out of parental care, they invariably do so, even in hunter-gatherer societies. Among Baka pygmies, for example, males who are good hunters contribute much less to the care of their offspring because they know that they are attractive bets for the ladies even when they don't do much at home. As with the monogamous monkeys and apes, it is likely that paternal care evolved *after* males became pairbonded because it becomes worth a pairbonded male's while making some contribution to the business of rearing in order to gain added benefits in terms of rearing success and/or offspring quality.

The bottom line here is that if two adults really are needed to rear a modern human baby, a much more plausible case can be made for grandmothers fulfilling this role, as Kristen Hawkes argued. Grandmothers are heavily involved in helping with the rearing of their daughters' children in all cultures, including our own. Again, there is some variation, but broadly speaking the evidence is clear that the mother's mother contributes more parental care than either the father's mother or the father – providing she doesn't live too far away. Even in contemporary industrialized societies, women naturally seem to exhibit a switch of interest from their spouses to their daughter(s) (in particular) at around the age of 45–50 just as the latter start to reproduce, as we were able to demonstrate from an analysis of telephone-call patterns. That grandmothering may be a very late evolutionary development is suggested by the fact that the fossil evidence clearly shows that it is only with the appearance of anatomically modern humans that a significant number of individuals survived to an age where they could be grandparents. From the australopithecines through to the Neanderthals, women may not have survived long enough to be grandmothers with any great frequency.

It has been suggested that grandmothering is the explanation for why human females – uniquely among mammals – undergo menopause (the termination of the capacity for reproduction around 45–50 years of age). Notwithstanding some rather unconvincing attempts to claim that other long-lived species (such as elephants and chimpanzees) also undergo menopause, a complete cessation of reproduction half way through the adult lifespan is unique to our species.[10]

If women are not using the menopause to switch from their own reproduction to helping daughters reproduce, then they are certainly doing so to see their own last offspring through to adulthood – a reflection of the fact that human parental investment is both expensive and *very* long term, and that offspring remain vulnerable if the parent(s) die prematurely.

So it seems that pairbonded relationships have benefits for both sexes: they reduce the risk of harassment for females and they allow males to monopolize matings with at least one female in a context where defending more females is impossible – due not to the females being too dispersed to be able to defend more than one but to their being at significantly increased risk of being poached by rivals. And there is the ever-present risk of infanticide. In a wide range of hunter-gatherer societies, men who acquire other men's wives (perhaps because the husband died or left the group) will kill the woman's young children. Even in our societies, step-children are more likely to be discriminated against than biological children, and are subject to significantly higher risk of being abused or murdered, especially by step-fathers.

When did pairbonding evolve?

This leaves us with one final question: exactly when did pairbonds evolve in the hominin lineage? As we have already seen, there have been several attempts to claim that monogamy (or should it really be pairbonding?) evolved extremely early, perhaps even as early as the australopithecines. However, as we concluded in Chapter 4, it seems very unlikely that the australopithecines had any kind of monogamous mating system.

Given our conclusion that monogamy (and pairbonding) is likely to have arisen in the context of infanticide risk, can we identify a point at which this risk became sufficiently large to warrant a switch from polygamy? We can get some handle on the level of infanticide risk in hominins by considering the likely size of foraging groups since the presence of several males is likely to exacerbate the problem. In the series of mathematical models that I developed for ape and feral goat mating strategies, the number of females foraging together was a key parameter influencing males' decisions on whether to be social (i.e. stay with a female group once they had found one) or opt for a form of roving-male polygyny (where they moved on to find a new group of fertile females once they had mated with the females in a given group). This relationship is also influenced by the typical length of the female's reproductive cycle (the interval between successive births), the density of female groups in the habitat and the male's search patterns (the area that a male carves out each day as he forages), but the key factor is how patchily the females are distributed.

Fig. 9.5 plots the proportion of males who are social (i.e. are with a female group at any given moment) against the ratio of the payoffs for the two strategies (being social versus being a roving male) for various great ape populations and one human hunter-gatherer population, as calculated from this mating strategy model. In species like chimpanzees and orang utans, where females are found in very small foraging groups or forage on their own, it doesn't pay males to

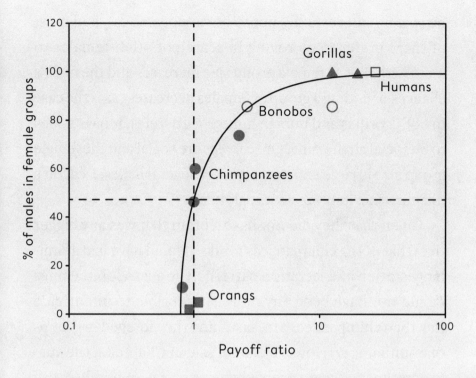

Figure 9.5

A model of male mating strategies for apes with two alternative strategies (being social and staying permanently with a group of females versus pursuing a roving-male strategy). The proportion of males that associate with groups of females in individual populations is plotted against the ratio of payoffs for the two strategies. The payoff for a social male is simply the number of females in the group that he stays with (in effect, the mean size of female groups); the payoff to a roving male is determined by the area the male can search each day, the size and density of female groups, and the likelihood that any given female will be sexually receptive on the day the male finds her group. When the payoff ratio is 1, the two strategies are equal; when the ratio is >1, it pays to be social, when <1 it pays to go roving. Notice that the data pass right through the midpoint (males should be ambivalent about whether to stay or leave a female group when the payoffs are equal), just as it should do in an optimal foraging model. The human data point is based on data from a !Kung San population. Full details of the model are given in Dunbar (1998).
Redrawn from Dunbar (2000).

be social (and certainly not to be monogamous), and most of them prefer to go roving in search of other females to mate with. But as female group size increases and the male's chances of finding a group of females decreases (as is the case in both gorillas and human hunter-gatherers), it pays males to be social all the time. Once males are social and the female group size is large enough to attract several males, Deacon's dilemma inevitably rears its ugly head.

Given that they did not have a brain that was any bigger than that of the chimpanzees (and so would have had a similar gestation and lactation pattern), the australopithecines would not have been any more susceptible to infanticide risk than chimpanzees are, assuming they foraged in groups of a similar size. However, as we saw in Chapter 4, the australopithecines were probably more social, in part because of higher predation risk away from the forests and in part because they occupied rich lakeside or riverine habitats that allowed them to forage in larger groups as a defence against predators. If this resulted in the females living in groups that were about ten times the size of those in which chimpanzee females typically live now,[11] then the males would certainly have been social and stayed with the female groups, just as we see in bonobo populations where many females forage together (Fig. 9.5). That would undoubtedly have increased the risk of harassment and infanticide. However, there is a world of difference between being social and being monogamous. Given the level of sexual dimorphism in the australopithecines (see Chapter 4), a promiscuous or harem-based mating strategy is the most likely, with the latter being the option of choice if infanticide risk was high.

With the appearance of early *Homo*, there is a modest increase in community size and in all likelihood, given the change to a more nomadic lifestyle, an increase in the size of the foraging group if only to provide protection from predators. But this increase was probably not enormous and a significant change in infanticide risk seems unlikely even with the modest change in gestation length that their larger brains would have demanded. By far the biggest change is likely to have come with the appearance of the *heidelbergs*: the 50 per cent increase in community size (and hence the number of males) would have dramatically increased the threat to females from harassment and infanticide, and this would have been exacerbated by the significantly longer interbirth intervals they experienced consequent on their significantly larger brains. At this point, a form of 'hired gun' strategy in which females attached themselves to a male for protection may well have had very significant advantages for the females. Even among modern humans, the risk of harassment by males in public places is radically reduced for single women if they are accompanied by a male (i.e. a bodyguard). However, as we saw in Chapter 6, the anatomical and genetic data both suggest that, if anything, archaic humans were polygamous, perhaps with a form of gorilla-like harem.[12] If the males are not providing much parental care (and, as we saw earlier, there is no anatomical evidence to suggest they were), then females will have no objection to sharing the protective benefits of the same male.

This problem would only have got worse with the appearance of modern humans, given a significant further increase in community (and hence camp group) size. With more males competing with each other to monopolize access to females, this is likely to have increased the pressure on dominant males. Bogusław Pawłowski and I showed some years ago that, in promiscuously mating primates, males find it impossible to monopolize mating access to the females in their group if there are more than four other males in the group: while they are busy mating with one female, their rivals can steal matings with the others. In such circumstances, males switch from trying to monopolize access to all the females to a one-mate-at-a-time form of promiscuity.[13] In other words, temporary pairbonds (as reflected in the 'consortships' of baboons and chimpanzees) emerge naturally as the number of rival males in the immediate vicinity increases and males are forced to guard individual females. If females bond to a male on a more permanent basis, this at least provides a transition into more permanent pairbonds as males in turn become increasingly willing to focus their social attention on this female beyond the immediate confines of her oestrous period in the interests of protecting her offspring. The decoupling of sexual behaviour from the menstrual cycle, as has happened in human females (and to a lesser extent only in bonobos and marmosets among the primates), probably played an important role in facilitating this by making it possible for pairs to have sex more or less continuously.

Fig. 4.5 plotted 2D:4D ratios for a selection of hominins, with comparison data from living apes and humans. There are no data for early *Homo*, but the five Neanderthals and one archaic human in the sample all had digit ratios at the lower end of the modern human range, not unlike those for gorillas with their harem-based mating system. None of these species, including modern humans, has a digit ratio commensurate with the kind of obligate lifelong monogamy that we find in gibbons. Taken together, this suggests that hominin mating systems were largely polygamous (i.e. promiscuous or harem-based), with probably only modern humans exhibiting any significant form of reciprocated pairbonding.

Some confirmatory evidence for polygamy in Neanderthals comes from the analysis of the mtDNA of a set of 12 Neanderthals from the El Sidrón site in northern Spain, who are thought to represent a single deposition event. The three males in the sample all belong to the same mtDNA lineage, whereas the three mature females all belong to completely different lineages. This suggests a patrilocal residence system similar to that found in all great apes, with females attracted in from neighbouring groups to live with a group of related males. Since, in modern humans, patrilocal residence systems are associated with polygamy whereas monogamy is always associated with bilateral dispersal, polygamy would seem a safe bet for the Neanderthal mating system. This would fit with the 2D:4D evidence, which suggests that Neanderthals fall at the polygamous end of the modern human distribution, roughly in line with the gorilla average.

The ambiguity in the modern human data on all the anatomical indices may reflect the fact that human mating strategies are in fact polymorphic. In other words, humans don't all follow the same strategy. Behaviourally, this is known to evolutionary psychologists as the 'cads vs dads' distinction. In any given population, some of the males are relatively monogamous and invest in their children (the dads), and others are more promiscuous and don't (the cads). This was illustrated rather nicely some years ago in a study of Québécois men by Daniel Pérusse. He found that the men divided into two behavioural types: a group that were more or less monogamous (at least in long term relationships) and a group of philanderers, in a ratio of approximately 2:1.[14] A recent Swedish twin study that examined men's vasopressin receptor genes in relation to their pairbondedness suggested that about 25 per cent of males had an allele for this gene that inclined them to behave more promiscuously. Similarly, in an analysis of both the 2D:4D ratio and a psychological measure of sociosexual preferences,[15] Rafael Wlodarski and I found that males split roughly 45:55 into monogamous and promiscuous types, with women showing a similar, but reversed, split. Although the ratio of the two phenotypes differs somewhat between samples, these data broadly support the suggestion that humans (of both sexes, it seems!) are divided into two sociosexual phenotypes: those that do and those that do not invest in long-term relationships (even though all of them may experience romantic attachment). More importantly, the twin study and the 2D:4D data suggest that these phenotypes have a genetic basis, even if the extent to which they are expressed can be modified to some extent by cultural influences.

Putting all this together, there is strong evidence to suggest hominins have been characterized by a polygamous ape-like mating system throughout most of their evolutionary history. If monogamy has evolved at all, it has probably done so only in modern humans, and even then it may not be full-blown monogamy *or* characteristic of the whole species. In humans, both sexes are polymorphic, with monogamous vs promiscuous phenotypes being close to an even split. If we take the contrast between gorillas and chimpanzees (plus orangs) as the benchmark for the difference between pairbonded and promiscuous mating relationships, the data would broadly seem to suggest some form of gorilla-like pseudo-pairbonded polygyny from at least archaic humans onwards (possibly even early *Homo*), giving way to a somewhat more strongly pair-bonded mating system in modern humans (but falling some way short of obligate monogamy).

Our odyssey through the story of human evolution has focused on the cognitive and social aspects of the story, rather than the stones and bones favoured by more conventional accounts. This is because, in the final analysis, the question of how we eventually came to be human is all about the social and cognitive traits that really differentiate us from the other apes. In these terms, the australopithecines that form the first phase of this story are really no more than transitional bipedal apes. The evolution of modern humans from this ancient ape stock has been a long road, absorbing the entire 2 million years from the appearance of early *Homo*. Although bipedalism obviously plays an important role in the rise of modern humans (it made it possible to gain control over breathing in a way that later turned out to be crucial for, first, laughter and, later, speech), this alone does not really warrant our viewing the australopithecines as more than great apes. They were a wonderful evolutionary experiment that, like all ecological radiations, tried many different variants around a theme, some very successfully (one, after all, eventually gave rise to us), others less so.

The real story of how we came to be who we are begins with the appearance of the first *Homo* species, conventionally *Homo ergaster* but arguably some of their immediate and more transient precursors (*Homo rudolfensis* and *H. gautengensis*). From there on in, it was a constant battle with time budgets under pressure from environmental factors that were selecting for ever larger community sizes – initially, perhaps, the need to provide protection from predators, but later the need to maintain trading networks for access to ephemeral resources, transmuting later still into the need for defence against conspecific raiding.

The story of human evolution has been one of finding ways to adjust to that pressure through novel solutions to the problems of social bonding and the nutrient demands of large bodies and brains. In some cases, we can see these as a series of stages (there have sometimes been quite dramatic phase shifts that resulted in very rapid evolutionary change), but in others as a slow steady change in a particular direction. But in the end, what has made us who we are has been a complex series of adjustments to the basic hominin physiological, social and cognitive design. It has, of course, been the cognitive changes that have given us the modern world of science and the arts, but it is the three together that have given us the rich tapestry of modern human relationships.

Bibliography

CHAPTER 1: WHAT WE HAVE TO EXPLAIN

- Balter, V., Braga, J., Télouk, P., and Thackeray, J. F. Evidence for dietary change but not landscape use in South African early hominins. *Nature* 489: 558–60.
- Brunet, M., Guy, F., Pilbeam, D., Mackaye, H. et al. (2002). A new hominid from the Upper Miocene of Chad, Central Africa. *Nature* 418: 145–51.
- De Miguel, C., and Heneberg, M. (2001). Variation in hominin brain size: how much is due to method? *Homo* 52: 3–58.
- Dunbar, R. I. M. (1993). Coevolution of neocortex size, group size and language in humans. *Behavioral and Brain Sciences* 16: 681–735.
- Dunbar, R. I. M. (2004). *The Human Story*. London: Faber and Faber.
- Dunbar, R. I. M. (2008). Mind the gap: or why humans aren't just great apes. *Proceedings of the British Academy* 154: 403–23.
- Dunbar, R. I. M., and Shultz, S. (2007). Understanding primate brain evolution. *Philosophical Transactions of the Royal Society, London* 362B: 649–58.
- Gowlett, J. A. J., Gamble, C., and Dunbar, R. I. M. (2012). Human evolution and the archaeology of the social brain. *Current Anthropology* 53: 693–722.
- Harrison, T. (2010). Apes among the tangled branches of human origins. *Science* 327: 532–4.
- Haslam, M., Hernandez-Aguílar, A., Ling, V. et al. (2009). Primate archaeology. *Nature* 460: 339–444.
- Ingman, M., Kaessmann, H., Pääbo, S., and Gyllensten, U. (2000). Mitochondrial genome variation and the origin of modern humans. *Nature* 408: 708–13.
- Klein, R. (1999). *The Human Career*, 2nd edition. Chicago: University of Chicago Press.
- Krause, J., Fu, Q., Good, J. et al. (2010). The complete mitochondrial DNA genome of an unknown hominin from southern Siberia. *Nature* 464: 894–97.

– Lahr, M. M., and Foley, R. (1994). Multiple dispersals and modern human origins. *Evolutionary Anthropology* 3: 48–60.

– Lockwood, C. A., Kimbel, W. H., and Lynch, J. M. (2004). Morphometrics and hominoid phylogeny: support for a chimpanzee–human clade and differentiation among great ape subspecies. *Proceedings of the National Academy of Sciences, USA* 101: 4356–60.

– McGrew, W. C. (1992). *Chimpanzee Material Culture: Implications for Human Evolution.* Cambridge: Cambridge University Press.

– Relethford, J. H. (1995). Genetics and modern human origins. *Evolutionary Anthropology* 4: 53–63.

– Reno, P., Meindl, R., McCollum, M., and Lovejoy, O. (2003). Sexual dimorphism in *Australopithecus afarensis* was similar to that of modern humans. *Proceedings of the National Academy of Sciences, USA* 100: 9404–9.

– Ruvolo, M. (1997). Molecular phylogeny of the hominoids: inferences from multiple independent DNA sequence data sets. *Molecular Biology and Evolution* 14: 248–65.

– Satta, Y., Klein, J., and Takahata, N. (2000). DNA archives and our nearest relative: the trichotomy problem revisited. *Molecular Phylogenetics and Evolution* 14: 259–75.

– Senut, B., Pickford, M., Gommery, D., Mein, P., Cheboi, K., and Coppens, Y. (2001). First hominid from the Miocene (Lukeino Formation, Kenya). *Comptes Rendus* 332: 137–44.

– Shultz, S., Nelson, E., and Dunbar, R. I. M. (2012). Hominin cognitive evolution: identifying patterns and processes in the fossil and archaeological record. *Philosophical Transactions of the Royal Society, London* 367B: 2130–40.

– Steudel-Numbers, K. L. (2006). Energetics in Homo erectus and other early hominins: the consequences of increased lower-limb length. *Journal of Human Evolution* 51: 445–53.

– Stoneking, M. (1993). DNA and recent human evolution. *Evolutionary Anthropology* 2: 60–73.

– Swedell, L., and Plummer, T. (2012). Papionin multilevel society as a model for hominin social evolution. *International Journal of Primatology* 33: 1165–93.

– Tooby, J., and DeVore, I. (1987). The reconstruction of hominid behavioural evolution through strategic modelling. In: W. G. Kinzey (ed.) *The Evolution of Human Behavior: Primate Models*, pp. 183–238. New York: State University of New York Press.

– Whiten, A., and Byrne, R. W. (eds) (1988). *Machiavellian Intelligence.* Oxford: Oxford University Press.

– Whiten, A., Horner, V., and Marshall-Pescini, S. (2003). Cultural panthropology. *Evolutionary Anthropology* 12: 92–105.
– Wynn, T., and Coolidge, F. L. (2004). The expert Neanderthal mind. *Journal of Human Evolution* 46: 467–87.

CHAPTER 2: THE BASES OF PRIMATE SOCIALITY

– Abbott, D. H., Keverne, E. B., Moore, G. F., and Yodyinguad, U. (1986). Social suppression of reproduction in subordinate talapoin monkeys, *Miopithecus talapoin*. In: J. Else and P. C. Lee (eds) *Primate Ontogeny*, pp. 329–41. Cambridge: Cambridge University Press.
– Altmann, J. (1980). *Baboon Mothers and Infants*. Cambridge, MA: Harvard University Press.
– Apperly, I. A. (2012). What is 'theory of mind'? Concepts, cognitive processes and individual differences. *Quarterly Journal of Experimental Psychology* 65: 825–39.
– Aron, A., Aron, E. N., and Smollan, D. (1992). Inclusion of other in the self scale and the structure of interpersonal closeness. *Journal of Personality and Social Psychology* 63: 596–612.
– Berscheid, E. (1994). Interpersonal relationships. *Annual Review of Psychology* 45: 79–129.
– Berscheid, E., Snyder, M., and Omoto, A. M. (1989). The relationship closeness inventory: assessing the closeness of interpersonal relationships. *Journal of Personality and Social Psychology* 57: 792–807.
– Bettridge, C., and Dunbar, R. I. M. (2012). Perceived risk and predation in primates: predicting minimum permissible group size. *Folia Primatologica* 83: 332–52.
– Bowman, L. A., Dilley, S. R., and Keverne, E. B. (1978). Suppression of oestrogen-induced LH surges by social subordination in talapoin monkeys. *Nature* 275: 56–8.
– Broad, K. D., Curley, J. P., and Keverne, E. B. (2006). Mother–infant bonding and the evolution of mammalian social relationships. *Philosophical Transactions of the Royal Society, London* 361B: 2199–214.
– Carrington, S. J., and Bailey, A. J. (2009). Are there Theory of Mind regions in the brain? A review of the neuroimaging literature. *Human Brain Mapping* 30: 2313–35.
– Cartmill, E. A., and Byrne, R. B. (2007). Orangutans modify their gestural signaling according to their audience's comprehension. *Current Biology* 17: 1–4.

– Cowlishaw, G. (1994). Vulnerability to predation in baboon populations. *Behaviour* 131: 293–304.
– Crockford, C., Wittig, R. M., Mundry, R., and Zuberbühler, K. (2012). Wild chimpanzees inform ignorant group members of danger. *Current Biology* 22: 142–6.
– Curly, J. P., and Keverne, E. B. (2005). Genes, brains and mammal social bonds. *Trends in Ecology and Evolution* 20: 561–7.
– Depue, R. A., and Morrone-Strupinsky, J. V. (2005). A neurobehavioral model of affiliative bonding: implications for conceptualizing a human trait of affiliation. *Behavioral and Brain Sciences* 28: 313–95.
– Dunbar, R. I. M. (1980). Determinants and evolutionary consequences of dominance among female gelada baboons. *Behavioral Ecology and Sociobiology* 7: 253–65.
– Dunbar, R. I. M. (1988). *Primate Social Systems*. London: Chapman & Hall.
– Dunbar, R. I. M. (1988). Habitat quality, population dynamics and group composition in colobus monkeys (*Colobus guereza*). *International Journal of Primatology* 9: 299–329.
– Dunbar, R. I. M. (1989). Reproductive strategies of female gelada baboons. In: A. Rasa, C. Vogel and E. Voland (eds) *Sociobiology of Sexual and Reproductive Strategies*, pp. 74–92. London: Chapman & Hall.
– Dunbar, R. I. M. (1991). Functional significance of social grooming in primates. *Folia Primatologica* 57: 121–31.
– Dunbar, R. I. M. (1995). The mating system of Callitrichid primates. I. Conditions for the coevolution of pairbonding and twinning. *Animal Behaviour* 50: 1057–70.
– Dunbar, R. I. M. (2010). Brain and behaviour in primate evolution. In: P. M. Kappeler and J. Silk (eds) *Mind the Gap: Tracing the Origins of Human Universals*, pp. 315–30. Berlin: Springer.
– Dunbar, R. I. M. (2010). The social role of touch in humans and primates: behavioural function and neurobiological mechanisms. *Neuroscience and Biobehavioral Reviews* 34: 260–68.
– Dunbar, R. I. M., and Dunbar, P. (1988). Maternal time budgets of gelada baboons. *Animal Behaviour* 36: 970–80.
– Dunbar, R. I. M., and Lehmann, J. (2013) Grooming and cohesion in primates: a comment on Grueter et al. *Evolution and Human Behavior* 34: 453–455.
– Dunbar, R. I. M., and Shultz, S. (2010). Bondedness and sociality. *Behaviour* 147: 775–803.
– Fedurek, P., and Dunbar, R. I. M. (2009). What does mutual grooming tell us about why chimpanzees groom? *Ethology* 115: 566–75.

– Gallagher, H. L., and Frith, C. D. (2003). Functional imaging of 'theory of mind'. *Trends in Cognitive Sciences* 7: 77–83.

– Granovetter, M. (1973). The strength of weak ties. *American Journal of Sociology* 78: 1360–80.

– Granovetter, M. (1983). The strength of weak ties: a network theory revisited. *Sociological Theory* 1: 201–33.

– Grueter, C. C., Bissonnette, A., Isler, K., and van Schaik, C. P. (2013). Grooming and group cohesion in primates: implications for the evolution of language. *Evolution and Human Behavior* 34: 61–8.

– Harcourt, A. H. (1992). Coalitions and alliances: are primates more complex than non-primates? In: A. H. Harcourt and F. B. M. de Waal (eds.) *Coalitions and Alliances in Humans and Other Animals*, pp. 445–72. Oxford: Oxford University Press.

– Harcourt, A. H., and Greenberg, J. (2001). Do gorilla females join males to avoid infanticide? A quantitative model. *Animal Behaviour* 62: 905–15.

– Hare, B., Call, J., Agnetta, B., and Tomasello, M. (2000). Chimpanzees know what conspecifics do and do not see. *Animal Behaviour* 59: 771–85.

– Hare, B., Call, J., and Tomasello, M. (2001). Do chimpanzees know what conspecifics know? *Animal Behaviour* 61: 139–51.

– Hill, R. A., and Dunbar, R. I. M. (1998). An evaluation of the roles of predation rate and predation risk as selective pressures on primate grouping behaviour. *Behaviour* 135: 411–30.

– Hill, R. A., and Lee, P. C. (1998). Predation pressure as an influence on group size in Cercopithecoid primates: implications for social structure. *Journal of Zoology* 245: 447–56.

– Hill, R. A., Lycett, J., and Dunbar, R. I. M. (2000). Ecological determinants of birth intervals in baboons. *Behavioral Ecolology* 11: 560–64.

– Huelsenbeck, J. P., Ronquist, F., Nielsen, R., and Bollback, J. P. (2001). Bayesian inference of phylogeny and its impact on evolutionary biology. *Science* 294: 2310–14.

– Isler, K., and van Schaik, C. P. (2006). Metabolic costs of brain size evolution. *Biology Letters* 2: 557–60.

– Karbowski, J. (2007). Global and regional brain metabolic scaling and its functional consequences. *BMC Biology* 5: 18–46.

– Keverne, E. B., Martensz, N., and Tuite, B. (1989). Beta-endorphin concentrations in cerebrospinal fluid of monkeys are influenced by grooming relationships. *Psychoneuroendocrinology* 14: 155–61.

– Kinderman, P., Dunbar, R. I. M., and Bentall, R. P. (1998). Theory-of-mind deficits and causal attributions. *British Journal of Psychology* 89: 191–204.

– Komers, P. E., and Brotherton, P. N. M. (1997). Female space use is the best predictor of monogamy in mammals. *Proceedings of the Royal Society, London* 264B: 1261–70.

– Lehmann, J., Korstjens, A. H., and Dunbar, R. I. M. (2007). Group size, grooming and social cohesion in primates. *Animal Behaviour* 74: 1617–29.

– Lewis, P. A., Birch, A., Hall, A., and Dunbar, R. I. M. (2013). Higher order intentionality tasks are cognitively more demanding: evidence for the social brain hypothesis.

– Lewis, P. A., Rezaie, R., Browne, R., Roberts, N., and Dunbar, R. I. M. (2011). Ventromedial prefrontal volume predicts understanding of others and social network size. *NeuroImage* 57: 1624–9.

– Machin, A., and Dunbar, R. I. M. (2011). The brain opioid theory of social attachment: a review of the evidence. *Behaviour* 148: 985–1025.

– O'Connell, S., and Dunbar, R. I. M. (2003). A test for comprehension of false belief in chimpanzees. *Evolution and Cognition* 9: 131–9.

– Opie, C., Atkinson, Q., Dunbar, R. I. M., and Shultz, S. (2013). Male infanticide leads to social monogamy in primates. *Proceedings of the National Academy of Sciences, USA* 110: 13328–32.

– van Overwalle, F. (2009). Social cognition and the brain: a meta-analysis. *Human Brain Mapping* 30: 829–58.

– Powell, J., Lewis, P. A., Dunbar, R. I. M., García-Fiñana, M., and Roberts, N. (2010). Orbital prefrontal cortex volume correlates with social cognitive competence. *Neuropsychologia* 48: 3554–62.

– Roberts, S. B. G., and Dunbar, R. I. M. (2011). The costs of family and friends: an 18-month longitudinal study of relationship maintenance and decay. *Evolution and Human Behavior* 32: 186–97.

– Roberts, S. B. G., Arrow, H., Lehmann, J., and Dunbar, R. I. M. (2014). Close social relationships: an evolutionary perspective. In: R. I. M. Dunbar, C. Gamble and J. A. J. Gowlett (eds) *Lucy to Language: The Benchmark Papers*, pp. 151–80. Oxford: Oxford University Press.

– van Schaik, C. P., and Dunbar, R. I. M. (1990). The evolution of monogamy in large primates: a new hypothesis and some crucial tests. *Behaviour* 115: 30–61.

– van Schaik, C. P., and Kappeler, P. M. (2003). The evolution of social monogamy in primates. In: Reichard, U. H., and Boesch, C. (eds) *Monogamy: Mating Strategies and Partnerships in Birds, Humans and Other Mammals*, pp. 59–80. Cambridge: Cambridge University Press.

– Shultz, S., Opie, C., and Atkinson, Q. D. (2011). Stepwise evolution of stable sociality in primates. *Nature* 479: 219–222.

– Silk, J. B., Alberts, S. C., and Altmann, J. (2003). Social bonds of female baboons enhance infant survival. *Science* 302: 1232–4.

– Silk, J. B., Beehner, J. C., Bergman, T. J., et al. (2009). The benefits of social capital: close social bonds among female baboons enhance offspring survival. *Proceedings of the Royal Society, London* 276B: 3099–104.

– Stiller, J., and Dunbar, R. I. M. (2007). Perspective-taking and memory capacity predict social network size. *Social Networks* 29: 93–104.

– Sutcliffe, A., Dunbar, R. I. M., Binder, J., and Arrow, H. (2012). Relationships and the social brain: integrating psychological and evolutionary perspectives. *British Journal of Psychology* 103: 149–68.

– Vrontou, S., Wong, A., Rau, K., Koerber, H., and Anderson, D. (2013). Genetic identification of C fibres that detect massage-like stroking of hairy skin in vivo. *Nature* 493: 669–73.

– Wittig, R. M., Crockford, C., Lehmann, J. et al. (2008). Focused grooming networks and stress alleviation in wild female baboons. *Hormones and Behavior* 54: 170–77.

CHAPTER 3: THE ESSENTIAL FRAMEWORK

– Barrickman, N. L., Bastian, M. L., Isler, K., and van Schaik, C. P. (2007). Life history costs and benefits of encephalization: a comparative test using data from long-term studies of primates in the wild. *Journal of Human Evolution* 54: 568–90.

– Barton, R. A., and Dunbar, R. I. M. (1997). Evolution of the social brain. In: A. Whiten and R. Byrne (eds) *Machiavellian Intelligence II*, pp. 240–63. Cambridge: Cambridge University Press.

– Bergman, T. J., Beehner, J. C., Cheney, D. L., and Seyfarth, R. M. (2003). Hierarchical classification by rank and kinship in baboons. *Science* 302: 1234–6.

– Bettridge, C., and Dunbar, R. I. M. (2013). Perceived risk and predation in primates: predicting minimum permissible group size. *Folia Primatologica* 83: 332–52.

– Bettridge, C., Lehmann, J., and Dunbar, R. I. M. (2010). Trade-offs between time, predation risk and life history, and their implications for biogeography: a systems modelling approach with a primate case study. *Ecological Modelling* 221: 777–90.

– Byrne, R. W., and Corp, N. (2004). Neocortex size predicts deception rate in primates. *Proceedings of the Royal Society, London* 271B: 1693–9.

– Curry, O., Roberts, S. B. G., and Dunbar, R. I. M. (2013). Altruism in social networks: evidence for a 'kinship premium'. *British Journal of Psychology* 104: 283–95.

– Deeley, Q., Daly, E., Asuma, R. et al. (2008). Changes in male brain responses to emotional faces from adolescence to middle age. *NeuroImage* 40: 389–97.

– Dunbar, R. I. M. (1988). *Primate Social Systems*. London: Chapman & Hall.

– Dunbar, R. I. M. (1992a). Neocortex size as a constraint on group size in primates. *Journal of Human Evolution* 22: 469–93.

– Dunbar, R. I. M. (1992b). A model of the gelada socio-ecological system. *Primates* 33: 69–83.

– Dunbar, R. I. M. (1993). Coevolution of neocortex size, group size and language in humans. *Behavioral and Brain Sciences* 16: 681–735.

– Dunbar, R. I. M. (1998). The social brain hypothesis. *Evolutionary Anthropology* 6: 178–90.

– Dunbar, R. I. M. (2008). Mind the gap: or why humans aren't just great apes. *Proceedings of the British Academy* 154: 403–23.

– Dunbar, R. I. M. (2011). Evolutionary basis of the social brain. In: J. Decety and J. Cacioppo (eds) *Oxford Handbook of Social Neuroscience*, pp. 28–38. Oxford: Oxford University Press.

– Dunbar, R. I. M. (2011). Constraints on the evolution of social institutions and their implications for information flow. *Journal of Institutional Economics* 7: 345–71.

– Dunbar, R. I. M., and Shi, J. (2013). Time as a constraint on the distribution of feral goats at high latitudes. *Oikos* 122: 403–10.

– Dunbar, R. I. M., and Shultz, S. (2007). Understanding primate brain evolution. *Philosophical Transactions of the Royal Society, London* 362B: 649–58.

– Dunbar, R. I. M., and Shultz, S. (2010). Bondedness and sociality. *Behaviour* 147: 775–803.

– Dunbar, R. I. M., Korstjens, A. H., and Lehmann, J. (2009). Time as an ecological constraint. *Biological Reviews of the Cambridge Philosophical Society* 84: 413–29.

– Elton, S. (2006). Forty years on and still going strong: the use of hominin–cercopithecid comparisons in palaeoanthropology. *Journal of the Royal Anthropological Institute* 12: 19–38.

– Fay, J. M., Carroll, R., Peterhans, J. C. K., and Harris, D. (1995). Leopard attack on and consumption of gorillas in the Central African Republic. *Journal of Human Evolution* 29: 93–9.

– Hamilton, M. J., Milne, B. T., Walker, R. S., Burger, O., and Brown, J. H. (2007). The complex structure of hunter-gatherer social networks. *Proceedings of the Royal Society, London* 274B: 2195–203.

– Hill, R. A., and Dunbar, R. I. M. (2003). Social network size in humans. *Human Nature* 14: 53–72.

– Hill, R. A., Bentley, A., and Dunbar, R. I. M. (2008). Network scaling reveals consistent fractal pattern in hierarchical mammalian societies. *Biology Letters* 4: 748–51.

– Joffe, T. H. (1997). Social pressures have selected for an extended juvenile period in primates. *Journal of Human Evolution* 32: 593–605.

– Joffe, T. H., and Dunbar, R. I. M. (1997). Visual and socio-cognitive information processing in primate brain evolution. *Proceedings of the Royal Society, London* 264B: 1303–7.

– Kanai, R., Bahrami, B., Roylance, R., and Rees, G. (2012). Online social network size is reflected in human brain structure. *Proceedings of the Royal Society, London* 279: 1327–34.

– Kelley, J. L., Morrell, L. J., Inskip, C., Krause, J., and Croft, D. P. (2011). Predation risk shapes social networks in fission-fusion populations. *PLoS-One* 6: e24280.

– Korstjens, A. H., and Dunbar, R. I. M. (2007). Time constraints limit group sizes and distribution in red and black-and-white colobus monkeys. *International Journal of Primatology* 28: 551–75.

– Korstjens, A. H., Lehmann, J., and Dunbar, R. I. M. (2010). Resting time as an ecological constraint on primate biogeography. *Animal Behaviour* 79: 361–74.

– Korstjens, A. H., Verhoeckx, I., and Dunbar, R. I. M. (2006). Time as a constraint on group size in spider monkey. *Behavioural Ecology and Sociobiology* 60: 683–94.

– Kudo, H., and Dunbar, R. I. M. (2001). Neocortex size and social network size in primates. *Animal Behaviour* 62: 711–22.

– Layton, R., O'Hara, S., and Bilsborough, A. (2012). Antiquity and social functions of multilevel social organization among human hunter-gatherers. *International Journal of Primatology* 33: 1215–45.

– Lehmann, J., and Dunbar, R. I. M. (2009). Network cohesion, group size and neocortex size in female-bonded Old World primates. *Proceedings of the Royal Society, London* 276B: 4417–22.

– Lehmann, J., and Dunbar, R. I. M. (2009). Implications of body mass and predation for ape social system and biogeographical distribution. *Oikos* 118: 379–90.

– Lehmann, J., Korstjens, A. H., and Dunbar, R. I. M. (2007). Group size, grooming and social cohesion in primates. *Animal Behaviour* 74: 1617–29.

– Lehmann, J., Korstjens, A. H., and Dunbar, R. I. M. (2007). Fission–fusion social systems as a strategy for coping with ecological constraints: a primate case. *Evolutionary Ecology* 21: 613–34.

– Lehmann, J., Korstjens, A. H., and Dunbar, R. I. M. (2008a). Time management in great apes: implications for gorilla biogeography. *Evolutionary Ecology Research* 10: 517–36.

– Lehmann, J., Korstjens, A. H., and Dunbar, R. I. M. (2008b). Time and distribution: a model of ape biogeography. *Ecology, Evolution and Ethology* 20: 337–59.

– Lehmann, J., Korstjens, A. H., and Dunbar, R. I. M. (2010). Apes in a changing world – the effects of global warming on the behaviour and distribution of African apes. *Journal of Biogeography* 37: 2217–31.

– Lehmann, J., Lee, P. C., and Dunbar, R. I. M. (2014). Unravelling the evolutionary function of communities. In: R. I. M. Dunbar, C. S. Gamble and J. A. J. Gowlett (eds) *Lucy to Language: The Benchmark Papers*, pp. 245–76. Oxford: Oxford University Press.

– Lewis, P. A., Rezaie, R., Browne, R., Roberts, N., and Dunbar, R. I. M. (2011). Ventromedial prefrontal volume predicts understanding of others and social network size. *NeuroImage* 57: 1624–9.

– Marlowe, F. G. (2005). Hunter-gatherers and human evolution. *Evolutionary Anthropology* 14: 54–67.

– Mink, J. W., Blumenschine, R. J., and Adams, D. B. (1981). Ratio of central nervous system to body metabolism in vertebrates – its constancy and functional basis. *American Journal of Physiology* 241: R203–12.

– O'Donnell, S., Clifford, M., and Molina, Y. (2011). Comparative analysis of constraints and caste differences in brain investment among social paper wasps. *Proceedings of the National Academy of Sciences, USA* 108: 7107–12.

– Palombit, R. A. (1999). Infanticide and the evolution of pairbonds in nonhuman primates. *Evolutionary Anthropology* 7: 117–29.

– Passingham, R. E., and Wise, S. P. (2012). *The Neurobiology of the Prefrontal Cortex*. Oxford: Oxford University Press.

– Pawłowski, B. P., Lowen, C. B., and Dunbar, R. I. M. (1998). Neocortex size, social skills and mating success in primates. *Behaviour* 135: 357–68.

– Pérez-Barbería, J., Shultz, S., and Dunbar, R. I. M. (2007). Evidence for intense coevolution of sociality and brain size in three orders of mammals. *Evolution* 61: 2811–21.

– Powell, J., Lewis, P. A., Roberts, N., García-Fiñana, M., and Dunbar, R. I. M. (2012). Orbital prefrontal cortex volume predicts social network size: an imaging study of individual differences in humans. *Proceedings of the Royal Society, London* 279B: 2157–62.

– de Ruiter, J., Weston, G., and Lyon, S. M. (2011). Dunbar's number: group size and brain physiology in humans reexamined. *American Anthropologist* 113: 557–68

– Roberts, S. B. G., and Dunbar, R. I. M. (2011). The costs of family and friends: an 18-month longitudinal study of relationship maintenance and decay. *Evolution and Human Behavior* 32: 186–97.

– Roberts, S. B. G., Dunbar, R. I. M., Pollet, T., and Kuppens, T. (2009). Exploring variations in active network size: constraints and ego characteristics. *Social Networks* 31: 138–46.

– Sallet, J., Mars, R. B., Noonan, M. P., et al. (2011). Social network size affects neural circuits in macaques. *Science* 334: 697–700.

– Saramäki, J., Leicht, E., López, E., Roberts, S., Reed-Tsochas, F., and Dunbar, R. I. M. (2014): The persistence of social signatures in human communication. *Proceedings of the National Academy of Sciences, USA.*

– Sayers, K., and Lovejoy, C. O. (2008). The chimpanzee has no clothes: a critical examination of *Pan troglodytes* in models of human evolution. *Current Anthropology* 49: 87–114.

– van Schaik, C. P. (1983). Why are diurnal primates living in groups? *Behaviour* 87: 91–117.

– Shultz, S., and Dunbar, R. I. M. (2006). Chimpanzee and felid diet composition is influenced by prey brain size. *Biology Letters* 2: 505–8.

– Shultz, S., and Dunbar, R. I. M. (2007). The evolution of the social brain: Anthropoid primates contrast with other vertebrates. *Proceedings of the Royal Society, London* 274B: 2429–36.

– Shultz, S., and Dunbar, R. I. M. (2010). Encephalisation is not a universal macroevolutionary phenomenon in mammals but is associated with sociality. *Proceedings of the National Academy of Sciences, USA* 107: 21582–6.

– Shultz, S., and Finlayson, L. V. (2010). Large body and small brain and group sizes are associated with predator preferences for mammalian prey. *Behavioral Ecology* 21: 1073–9.

– Shultz, S., Noe, R., McGraw, S., and Dunbar, R. I. M. (2004). A community-level evaluation of the impact of prey behavioural and ecological characteristics on predator diet composition. *Proceedings of the Royal Society, London* 271B: 725–32.

– Smith, A. R., Seid, M. A., Jimenez, L., and Wcislo, W. T. (2010). Socially induced brain development in a facultatively eusocial sweat bee *Megalopta genalis* (Halictidae). *Proceedings of the Royal Society, London* 277B: 2157–63.

– Smuts, B. B., and Nicholson, N. (1989). Dominance rank and reproduction in female baboons. *American Journal of Primatology* 19: 229–46.

– Tsukahara, T. (1993). Lions eat chimpanzees: the first evidence of predation by lions on wild chimpanzees. *American Journal of Primatology* 29: 1–11.

– Wellman, B. (2012). Is Dunbar's number up? *British Journal of Psychology.* 103: 174–6.

- Willems, E., and Hill, R. A. (2009). A critical assessment of two species distribution models taking vervet monkeys (*Cercopithecus aethiops*) as a case study. *Journal of Biogeography* 36: 2300–312.
- Zhou, W.-X., Sornette, D., Hill, R. A., and Dunbar, R. I. M. (2005). Discrete hierarchical organization of social group sizes. *Proceedings of the Royal Society, London* 272B: 439–44.

CHAPTER 4: THE FIRST TRANSITION: THE AUSTRALOPITHECINES

- Barrett, L., Gaynor, D., Rendall, D., Mitchell, D., and Henzi, S. P. (2004). Habitual cave use and thermoregulation in chacma baboons (*Papio hamadryas ursinus*). *Journal of Human Evolution* 46: 215–22.
- Boesch-Achermann, H., and Boesch, C. (1993). Tool use in wild chimpanzees: new light from dark forests. *Current Directions in Psychological Science* 2: 18–22.
- Boesch, C., and Boesch, H. (1983). Optimization of nut-cracking with natural hammers by wild chimpanzess. *Behaviour* 83: 265–86.
- Berger, L. (2007). Further evidence for eagle predation of, and feeding damage on, the Taung child. *South African Journal of Science* 103: 496–8.
- Bettridge, C. M. (2010). *Reconstructing Australopithecine Socioecology Using Strategic Modelling Based on Modern Primates*. DPhil thesis, University of Oxford.
- Brain, C. K. (1970). New finds at the Swartkrans australopithecine site. *Nature* 225: 1112–19.
- Carvalho, S., Biro, D., Cunha, E. et al. (2012). Chimpanzee carrying behaviour and the origins of human bipedality. *Current Biology* 22: R180–81.
- Cerling, T., Mbua, E., Kirera, F. et al. (2011). Diet of *Paranthropus boisei* in the early Pleistocene of East Africa. *Proceedings of the National Academy of Sciences, USA* 108: 9337–41.
- Copeland, S., Sponheimer, M., de Ruiter, J. et al. (2011). Strontium isotope evidence for landscape use by early hominins. *Nature* 474: 76–9.
- Dezecache, G., and Dunbar, R. I. M. (2012). Sharing the joke: the size of natural laughter groups. *Evolution and Human Behavior* 33: 775–9.
- Dunbar, R. I. M. (2010). Deacon's dilemma: the problem of pairbonding in human evolution. In: R. I. M. Dunbar, C. Gamble and J. A. G. Gowlett (eds) *Social Brain, Distributed Mind*, pp. 159–79. Oxford: Oxford University Press.
- Foley, R. A., and Elton, S. (1995). Time and energy: the ecological context for the evolution of bipedalism. In: E. Strasser, J. Fleagle, A. Rosenberger and H. McHenry (eds) *Primate Locomotion: Recent Advances*, pp. 419–33. New York: Plenum Press.

– Hunt, K. D. (1994). The evolution of human bipedality: ecology and functional morphology. *Journal of Human Evolution* 26: 183–202.

– Klein, R. G. (2000). *The Human Career: Human Biological and Cultural Origins*, 3rd edition. Chicago: Chicago University Press.

– Lawrence, K. T., Sosdian, S., White, H. E., and Rosenthal, Y. (2010). North Atlantic climate evolution through the Plio-Pleistocene climate transitions. *Earth and Planetary Science Letters* 300: 329–42.

– Lehmann, J., and Dunbar, R. I. M. (2009). Implications of body mass and predation for ape social system and biogeographical distribution. *Oikos* 118: 379–90.

– Lovejoy, C. O. (1981). The origin of man. *Science* 211: 341–50.

– Lovejoy, C. O. (2009). Reexamining human origins in light of *Ardipithecus ramidus. Science* 326: 74e1–8.

– McGraw, W. S., Cooke, C., and Shultz, S. (2006). Primate remains from African crowned eagle (*Stephanoaetus coronatus*) nests in Ivory Coast's Tai Forest: Implications for primate predation and early hominid taphonomy in South Africa. *American Journal of Physical Anthropology* 131: 151–65.

– McPherron, S., Alemseged, Z., Marean, C. et al. (2010). Evidence for stone-tool-assisted consumption of animal tissues before 3.39 million years ago at Dikika, Ethiopia. *Nature* 466: 857–60.

– Marlowe, F., and Berbesque, J. (2009). Tubers as fallback foods and their impact on Hadza hunter-gatherers. *American Journal of Physical Anthropology* 140: 751–58.

– Nelson, E., and Shultz, S. (2010). Finger length ratios (2D:4D) in anthropoids implicate reduced prenatal androgens in social bonding. *American Journal of Physical Anthropology* 141: 395–405.

– Nelson, E., Rolian, C., Cashmore, L., and Shultz, S. (2011). Digit ratios predict polygyny in early apes, *Ardipithecus*, Neanderthals and early modern humans but not in Australopithecus. *Proceedings of the Royal Society, London* 278B: 1556–63.

– Pawłowski, B. P., Lowen, C. B., and Dunbar, R. I. M. (1998). Neocortex size, social skills and mating success in primates. *Behaviour* 135: 357–68.

– Platt, J. R. (1964). Strong inference. *Science* 146: 347–53.

– Pontzer, H., Raichlen, D. A., Sockol, M. D. (2009). The metabolic costs of walking in humans, chimpanzees and early hominins. *Journal of Human Evolution* 56: 43–54.

– Reno, P. L., McCollum, M. A., Meindl, R. S., and Lovejoy, C. O. (2010). An enlarged postcranial sample confirms *Australopithecus afarensis* dimorphism was similar to modern humans. *Philosophical Transactions of the Royal Society* 365B: 3355–63.

– Reno, P. L., Meindl, R. S., McCollum, M. A., and Lovejoy, C. O. (2003). Sexual dimorphism in *Australopithecus afarensis* was similar to that of modern humans. *Proceedings of the National Academy of Sciences, USA* 100: 9404–9.
– Richmond, B. G., Aiello, L. C., and Wood, B. (2002). Early hominin limb proportions. *Journal of Human Evolution* 43: 529–48.
– Richmond, B. G., Strait, D. S., and Begun, D. R. (2001). Origin of human bipedalism: the knuckle-walking hypothesis revisited. *Yearbook of Physical Anthropology* 44: 70–105.
– Ruxton, G. D., and Wilkinson, D. M. (2011). Thermoregulation and endurance running in extinct hominins: Wheeler's models revisited. *Journal of Human Evolution* 61: 169–75.
– Ruxton, G. D., and Wilkinson, D. M. (2011). Avoidance of overheating and selection for both hair loss and bipedality in hominins. *Proceedings of the National Academy of Sciences, USA* 108: 20965–9.
– Schmid, P., Churchill, S. E., Nalla, S. et al. (2013). Mosaic morphology in the thorax of *Australopithecus sediba*. *Science* 340.
– Sockol, M. D., Raichlen, D. A., and Pontzer, H. (2007). Chimpanzee locomotor energetics and the origin of human bipedalism. *Proceedings of the National Academy of Sciences, USA* 104: 12265–9.
– Sponheimer, M., and Lee-Thorpe, J. (2003). Differential resource utilization by extant great apes and australopithecines: towards solving the C_4 conundrum. *Comparative Biochemistry and Physiology* 136A: 27–34.
– Sponheimer, M., Lee-Thorpe, J., de Ruiter, D. et al. (2005). Hominins, sedges, and termites: new carbon isotope data from the Sterkfontein valley and Kruger National Park. *Journal of Human Evolution* 48: 301–12.
– Tsukahara, T. (1993). Lions eat chimpanzees: the first evidence of predation by lions on wild chimpanzees. *American Journal of Primatology* 29: 1–11.
– Ungar, P. S., and Sponheimer, M. (2011). The diets of early hominins. *Science* 334: 190–93.
– Ungar, P. S., Grine, F. E., and Teaford, M. F. (2006). Diet in early *Homo*: a review of the evidence and a new model of adaptive versatility. *Annual Review of Anthropology* 35: 209–28.
– Wheeler, P. E. (1984). The evolution of bipedality and loss of functional body hair in hominids. *Journal of Human Evolution* 13: 91–8.
– Wheeler, P. E. (1985). The loss of functional body hair in man: the influence of thermal environment, body form and bipedality. *Journal of Human Evolution* 14: 23–8.
– Wheeler, P. E. (1991). The thermoregulatory advantages of hominid bipedalism in open equatorial environments: the contribution of increased

convective heat loss and cutaneous evaporative cooling. *Journal of Human Evolution* 21: 107–15.

- Wheeler, P. E. (1991). The influence of bipedalism on the energy and water budgets of early hominids. *Journal of Human Evolution* 21: 117–36.

- Wheeler, P. E. (1992). The thermoregulatory advantages of large body size for hominids foraging in savannah environments. *Journal of Human Evolution* 23: 351–62.

- Wheeler, P. E. (1992). The influence of the loss of functional hair on the water budgets of early hominids. *Journal of Human Evolution* 23: 379–88.

- Wheeler, P. E. (1993). The influence of stature and body form on hominid energy and water budgets: a comparison of Australopithecus and early *Homo* physiques. *Journal of Human Evolution* 24: 13–28.

CHAPTER 5: THE SECOND TRANSITION: EARLY *HOMO*

- Aiello, L. C., and Wells, J. (2002). Energetics and the evolution of the genus *Homo*. *Annual Review of Anthropology* 31: 323–38.

- Aiello, L. C., and Wheeler, P. (1995). The expensive tissue hypothesis: the brain and the digestive system in human evolution. *Current Anthropology* 36: 199–221.

- Allen, K. L., and Kay, R. F. (2012). Dietary quality and encephalization in platyrrhine primates. *Proceedings of the Royal Society, London* 279B: 715–21.

- Alperson-Afil, N. (2008). Continual fire-making by hominins at Gesher Benot Ya'aqov, Israel. *Quaternary Science Reviews* 27: 1733–9.

- Bailey, D., and Geary, D. (2009). Hominid Brain Evolution. *Human Nature* 20: 67–79.

- Barbetti, M., Clark, J. D., Williams, F. M., and Williams, M. A. J. (1980). Palaeomagnetism and the search for very ancient fireplaces in Africa. Results from a million-year-old Acheulian site in Ethiopia. *Anthropologie* 18: 299–304.

- Behrensmeyer, A. K., Todd, N. E., Potts, R., and McBrinn, G. E. (1997). Late Pliocene faunal turnover in the Turkana Basin, Kenya and Ethiopia. *Science* 278: 1589–94.

- Bellomo, R. V. (1994). Methods of determining early hominid behavioural activities associated with the controlled use of fire at FxJj20 Main, Koobi Fora, Kenya. *Journal of Human Evolution* 27: 173–95.

- Berna, F., Goldberg, P., Horwitz, L. K. et al. (2012). Microstratigraphic evidence of in situ fire in the Acheulean strata of Wonderwerk Cave, Northern Cape Province, South Africa. *Proceedings of the National Academy of Sciences, USA* 109: E1215–20.

- Binford, L. R., and Ho, C. K. (1985). Taphonomy at a distance: Zhoukoudian, 'the cave home of Beijing man'? *Current Anthropology* 26: 413–42.
- Brain, C. K., and Sillen, A. (1988). Evidence from the Swartkrans cave for the earliest use of fire. *Nature* 336: 464–6.
- Brown, K. S., Marean, C. W., Herries, A. I. R. et al. (2009). Fire as an engineering tool of early modern humans. *Science* 325: 859–62.
- Carmody, R. N., and Wrangham, R. W. (2009). The energetic significance of cooking. *Journal of Human Evolution* 57: 379–91.
- Clark, J. D., and Harris, J. W. K. (1985). Fire and its roles in early hominid lifeways. *African Archaeological Review* 3: 3–27.
- Coqueugniot, H., Hublin, J.-J., Veillon, F., Houët, F., and Jacob, T. (2004). Early brain growth in *Homo erectus* and implications for cognitive ability. *Nature* 431: 299–302.
- Cordain, L., Miller, J. B., Eaton, S. B., Mann, N., Holt, S. H. A., and Speth, J. D. (2000). Plant-animal subsistence ratios and macronutrient energy estimations in worldwide hunter-gatherer diets. *American Journal of Clinical Nutrition* 71: 682–92.
- Davila Ross, M., Allcock, B., Thomas, C., and Bard, K. A. (2011). Aping expressions? Chimpanzees produce distinct laugh types when responding to laughter of others. *Emotion* 11: 1013–20.
- Davila Ross, M., Owren, M. J., and Zimmermann, E. (2009). Reconstructing the evolution of laughter in great apes and humans. *Current Biology*, 19, 1–6.
- deMenocal, P. B. (2004). African climate change and faunal evolution during the Pliocene/Pleistocene. *Earth and Planetary Science Letters* 220: 3–24.
- De Miguel, C., and Heneberg, M. (2001). Variation in hominin brain size: how much is due to method? *Homo* 52: 3–58.
- Dezecache, G., and Dunbar, R. I. M. (2012). Sharing the joke: the size of natural laughter groups. *Evolution and Human Behavior* 33: 775–9.
- Dunbar, R. I. M. (2000). Male mating strategies: a modelling approach. In: P. Kappeler (ed.) *Primate Males*, pp. 259–68. Cambridge: Cambridge University Press.
- Dunbar, R. I. M. (2012). Bridging the bonding gap: the transition from primates to humans. *Philosophical Transactions of the Royal Society, London* 367B: 1837–46.
- Dunbar, R. I. M., and Gowlett, J. A. J. (2013). Fireside chat: the impact of fire on hominin socioecology. In: R. I. M. Dunbar, C. Gamble and J. A. J. Gowlett (eds) *The Lucy Project: The Benchmark Papers*, pp. 277–96. Oxford: Oxford University Press.
- Dunbar, R. I. M., and Shultz, S. (2007). Understanding primate brain evolution. *Philosophical Transactions of the Royal Society, London* 362B: 649–58.

- Dunbar, R. I. M., Baron, R., Frangou, A. et al. (2012). Social laughter is correlated with an elevated pain threshold. *Proceedings of the Royal Society, London* 279B: 1161–7.
- Dunbar, R. I. M., Marriot, A., and Duncan, N. (1997). Human conversational behaviour. *Human Nature* 8: 231–46.
- Gonzalez-Voyer, A., Winberg, S., and Kolm, N. (2009). Social fishes and single mothers: brain evolution in African cichlids. *Proceedings of the Royal Society, London* 276B: 161–7.
- Goren-Inbar N., Alperson N., Kislev, M. E. et al. (2004). Evidence of hominin control of fire at Gesher Benot Ya'aqov, Israel. *Science* 304: 725–7.
- Goudsbloom, J. (1995). *Fire and Civilisation*. Harmondsworth: Penguin.
- Gowlett, J. A. J. (2006). The early settlement of northern Europe: fire history in the context of climate change and the social brain. *Comptes Rendus Palevol* 5: 299–310
- Gowlett, J. A. J. (2010). Firing up the social brain. In: R. I. M. Dunbar, C. Gamble and J. A. J. Gowlett (eds) *Social Brain and Distributed Mind*, pp. 345–70. Oxford: Oxford University Press.
- Gowlett, J. A. J., and Wrangham, R. W. (2013). Earliest fire in Africa: towards convergence of archaeological evidence and the cooking hypothesis. *Azania* 48: 5–30.
- Gowlett, J. A. J., Hallos, J., Hounsell, S., Brant, V., and Debenham, N. C. (2005). Beeches Pit – archaeology, assemblage dynamics and early fire history of a Middle Pleistocene site in East Anglia, UK. *Eurasian Prehistory* 3: 3–38.
- Gowlett, J. A. J., Harris, J. W. K., Walton, D., and Wood, B. A. (1981). Early archaeological sites, hominid remains and traces of fire from Chesowanja, Kenya. *Nature* 294: 125–9.
- Hallos J. (2005). '15 Minutes of Fame': exploring the temporal dimension of Middle Pleistocene lithic technology. *Journal of Human Evolution* 49: 155–79.
- Hartwig, W., Rosenberger, A., Norconk, M., and Owl, M. (2011). Relative brain size, gut size, and evolution in New World Monkeys. *Anatomical Record* 294: 2207–21.
- Isler, K., and van Schaik, C. P. (2009). The expensive brain: a framework for explaining evolutionary changes in brain size. *Journal of Human Evolution* 57: 392–400.
- Isler, K., and van Schaik, C. P. (2012). How our ancestors broke through the gray ceiling. *Current Anthropology* 53: S453–65.
- Klein, R. G. (2000). *The Human Career: Human Biological and Cultural Origins*, 3rd edition. Chicago: Chicago University Press.

- Kotrschal, A., Rogell, B., Bundsen, A. et al. (2013). Artificial selection on relative brain size in the guppy reveals costs and benefits of evolving a larger brain. *Current Biology* 23: 1–4.
- Larson, S. G. (2007). Evolutionary transformation of the hominin shoulder. *Evolutionary Anthropology* 16: 172–87.
- Lehmann, J., Korstjens, A. H., and Dunbar, R. I. M. (2007). Group size, grooming and social cohesion in primates. *Animal Behaviour* 74: 1617–29.
- Leonard, W. R., Robertson, M. L., Snodgrass, J. J., and Kuzawa, C. W. (2003). Metabolic correlates of hominid brain evolution. *Comparative Biochemistry and Physiology* 136A: 5–15.
- Ludwig, B. (2000). New evidence for the possible use of controlled fire from ESA sites in the Olduvai and Turkana basins. *Journal of Human Evolution* 38: A17.
- de Lumley, H. (2006). Il y a 400,000 ans: la domestication du feu, un formidable moteur d'hominisation. In: H. de Lumley (ed.) *Climats, Cultures et Sociétés aux Temps Préhistoriques, de l'Apparition des Hominidés Jusqu'au Néolithique. Comptes Rendus Palevol* 5: 149–54.
- McKinney, C. (2001). The uranium-series age of wood from Kalambo Falls. Appendix D in: J. D. Clark (ed.) 2001. *Kalambo Falls*, Vol. 3, pp. 665–74. Cambridge: Cambridge University Press.
- Maslin, M. A., and Trauth, M. H. (2009). Plio-Pleistocene East African pulsed climate variability and its influence on early human evolution. In: F. E. Grine, J. G. Fleagle and R. E. Leakey (eds.) *The First Humans: Origin and Early Evolution of the Genus Homo*, pp. 151–8. Berlin: Springer.
- Morwood, M., Soejono, R., Roberts, R. et al. Archaeology and age of a new hominin from Flores in eastern Indonesia. *Nature* 431: 1087–91. http://www.nature.com/nature/journal/v431/n7012/abs/nature02956.html – a8(2004).
- Navarette, A., van Schailk, C. P., and Isler, K. (2011). Energetics and the evolution of human brain size. *Nature* 480: 91–3.
- Niven, J. E., and Laughlin, S. B. (2008). Energy limitation as a selective pressure on the evolution of sensory systems. *Journal of Experimental Biology* 211: 1792–804.
- Osaka City University (2011). Catalogue of Fossil Hominids Database. http://gbs.ur-plaza.osaka-cu.ac.jp/kaseki/index.html
- Pawtowski, B. P., Lowen, C. B., and Dunbar, R. I. M. (1998). Neocortex size, social skills and mating success in primates. *Behaviour* 135: 357–68.
- Plavcan, J. M. (2012). Body size, size variation, and sexual size dimorphism in early *Homo. Current Anthropology* 53: S409–23.
- Preece, R. C., Gowlett, J. A. J., Parfitt, S. A., Bridgland, D. R., and Lewis, S. G.

(2006). Humans in the Hoxnian: habitat, context and fire use at Beeches Pit, West Stow, Suffolk, UK. *Journal of Quaternary Science* 21: 485–96.

– Provine, R. (2000). *Laughter*. Harmondsworth: Penguin Books.

– Richmond, B. G., Aiello, L. C., and Wood, B. (2002). Early hominin limb proportions. *Journal of Human Evolution* 43: 529–48.

– Roach, N. T., Venkadesan, M., Rainbow, M. J., and Lieberman, D. E. (2013). Elastic energy storage in the shoulder and the evolution of high-speed throwing in *Homo*. *Nature* 498: 483–7.

– Roebroeks, W., and Villa, P. (2011). On the earliest evidence for habitual use of fire in Europe. *Proceedings of the National Academy of Sciences, USA* 108: 5209–14.

– Rolland, N. (2004). Was the emergence of home bases and domestic fire a punctuated event? A review of the Middle Pleistocene record in Eurasia. *Asian Perspectives* 43: 248–80.

– Shipman, P., and Walker, A. (1989). The costs of becoming a predator. *Journal of Human Evolution* 18: 373–92.

– Shultz, S., and Dunbar, R. I. M. (2010). Social bonds in birds are associated with brain size and contingent on the correlated evolution of life-history and increased parental investment. *Biological Journal of the Linnean Society* 100: 111–23.

– Shultz, S., and Dunbar, R. I. M. (2010). Encephalisation is not a universal macroevolutionary phenomenon in mammals but is associated with sociality. *Proceedings of the National Academy of Sciences, USA* 107: 21582–6.

– Shultz, A., and Maslin, M. (2013). Early human speciation, brain expansion and dispersal influenced by African climate pulses. *PLoS One* 8: e76750.

– Simpson, S. W., Quade, J., Levin, N. E. et al. (2008). A female *Homo erectus* pelvis from Gona, Ethiopia. *Science* 322: 1089–92.

– Speth, J. D. (1991). Protein selection and avoidance strategies of contemporary and ancestral foragers: unresolved issues. *Philosophical Transactions of the Royal Society, London* 334: 265–70.

– Ungar, P. S. (2012). Dental evidence for the reconstruction of diet in African early *Homo*. *Current Anthropology* 53: S318–29.

– Weiner S., Xu Q., Goldberg P., Lui J., and Bar-Yosef, O. (1998). Evidence for the use of fire at Zhoukoudian, China. *Science* 281: 251–3.

– Williams, D. F., Peck, J., Karabanov, E. B. et al. (1997). Lake Baikal record of continental climate response to orbital insolation during the past 5 million years. *Science* 278: 1114–17.

– Wood, B., and Collard, M. (1999). The human genus. *Science* 284: 65–71.

- Wrangham, R. W. (2010). *Catching Fire: How Cooking Made Us Human*. New York: Basic Books.
- Wrangham, R. W., and Conklin-Brittain, N. (2003). Cooking as a biological trait. *Comparative Biochemistry and Physiology* A, 136: 35–46.
- Wrangham, R. W., and Peterson, D. (1996). *Demonic Males: Apes and the Origins of Human Violence*. New York: Houghton Mifflin.
- Wrangham, R. W., Jones, J. H., Laden, G., Pilbeam, D., and Conklin-Brittain, N. (1999). The raw and the stolen: cooking and the ecology of human origins. *Current Anthropology* 40: 567–94.
- Wrangham, R. W., Wilson, M. L., and Muller, M. N. (2006). Comparative rates of violence in chimpanzees and humans. *Primates* 47: 14–26.
- Wu, X., Schepartz, L. A., Falk, D., and Liu, W. (2006). Endocranial cast of Hexian *Homo erectus* from South China. *American Journal of Physical Anthropology* 130: 445–54.

CHAPTER 6: THE THIRD TRANSITION: ARCHAIC HUMANS

- Bailey, D., and Geary, D. (2009). Hominid brain evolution. *Human Nature* 20: 67–79.
- Beals, K. L., Courtland, L. S., Dodd, S. M. et al. (1984). Brain size, cranial morphology, climate, and time machines. *Current Anthropology* 25: 301–30.
- Bergman (2013). Speech-like vocalized lip-smacking in geladas. *Current Biology* 23: R268–9.
- Arsuaga, J. L., Bermúdez de Castro, J. M., and Carbonell, E. (eds.) (1997). The Sima de los Huesos hominid site. *Journal of Human Evolution* 33: 105–421.
- Balzeau, A., Holloway, R. L., and Grimaud-Hervé, D. (2012). Variations and asymmetries in regional brain surface in the genus *Homo*. *Journal of Human Evolution* 62: 696–706.
- Bruner, E., Manzi, G., and Arsuaga, J. L. (2003). Encephalization and allometric trajectories in the genus *Homo*: evidence from the Neandertal and modern lineages. *Proceedings of the National Academy of Sciences, USA* 100: 15335–40.
- Carbonell, E., and Mosquera, A. (2006). The emergence of a symbolic behaviour: the sepulchral pit of Sima de los Huesos, Sierra de Atapuerca, Burgos, Spain. *Comptes Rendus Palevol* 5: 155–60.
- Churchill, S. E. (1998). Cold adaptation, heterochrony, and Neandertals. *Evolutionary Anthropology* 7: 46–61.
- Cohen, E., Ejsmond-Frey, R., Knight, N., and Dunbar, R. I. M. (2010). Rowers'

high: behavioural synchrony is correlated with elevated pain thresholds. *Biology Letters* 6: 106–8.

– Dunbar, R. I. M. (2011). On the evolutionary function of song and dance. In: N. Bannan (ed.) *Music, Language and Human Evolution*, pp. 201–14. Oxford: Oxford University Press.

– Dunbar, R. I. M., and Shi, J. (2013). Time as a constraint on the distribution of feral goats at high latitudes. *Oikos* 122: 403–10.

– Dunbar, R. I. M., Kaskatis, K., MacDonald, I., and Barra, V. (2012). Performance of music elevates pain threshold and positive affect. *Evolutionary Psychology* 10: 688–702.

– Foley, R. A., and Lee, P. C. (1989). Finite social space, evolutionary pathways, and reconstructing hominid behavior. *Science* 243: 901–6.

– Gunz, P., Neubauer, S., Golovanova, L. et al. (2012). A uniquely modern human pattern of endocranial development. Insights from a new cranial reconstruction of the Neandertal newborn from Mezmaiskaya. *Journal of Human Evolution* 62: 300–13.

– Gustison, M. L., le Roux, A., and Bergman, T. J. (2012). Derived vocalizations of geladas (*Theropithecus gelada*) and the evolution of vocal complexity in primates. *Philosophical Transactions of the Royal Society, London* B 367B: 1847–59.

– Holmes, J. A., Atkinson, T., Darbyshire, D. P. F. et al. (2010). Middle Pleistocene climate and hydrological environment at the Boxgrove hominin site (West Sussex, UK) from ostracod records. *Quaternary Science Reviews* 29: 1515–27.

– Joffe, T., and Dunbar, R. I. M. (1997). Visual and socio-cognitive information processing in primate brain evolution. *Proceedings of the Royal Society, London* 264B: 1303–7.

– Kirk, E. C. (2006). Effects of activity pattern on eye size and orbital aperture size in primates. *Journal of Human Evolution* 51: 159–70.

– Klein, R. G. (2000). *The Human Career: Human Biological and Cultural Origins*, 3rd edition. Chicago: Chicago University Press.

– Krings, M., Stone, A., Schmitz, R. W., Krainitzki, H., Stoneking, M., and Pääbo, S. (1997). Neandertal DNA sequences and the origin of modern humans. *Cell* 90: 19–30.

– Lalueza-Fox, C., Rosas, A., Estalrrich, A. et al. (2010). Genetic evidence for patrilocal mating behaviour among Neandertal groups. *Proceedings of the National Academy of Sciences, USA* 108: 250–53.

– McNeill, W. H. (1995). *Keeping in Time Together: Dance and Drill in Human History*. Cambridge, MA: Harvard University Press.

– Maslin, M. A., and Trauth, M. H. (2009). Plio-Pleistocene East African pulsed climate variability and its influence on early human evolution. In: F. E. Grine, J. G. Fleagle and R. E. Leakey (eds) *The First Humans: Origin and Early Evolution of the Genus Homo*, pp. 151–8. Berlin: Springer.

– Mithen, S. (2005). *The Singing Neanderthals: The Origins of Music, Language, Mind and Body*. Cambridge, MA: Harvard University Press.

– Niven, L., Steele, T., Rendu, W. et al. (2012). Neandertal mobility and large-game hunting: the exploitation of reindeer during the Quina Mousterian at Chez-Pinaud Jonzac (Charente-Maritime, France). *Journal of Human Evolution* 63: 624–35.

– Noonan, J. P., Coop, G., Kudaravalli, S. et al. (2006). Sequencing and analysis of Neanderthal genomic DNA. *Science* 314: 1113–18.

– Osaka City University (2011). Catalogue of Fossil Hominids Database. http://gbs.ur-plaza.osaka-cu.ac.jp/kaseki/index.html.

– Pearce, E., and Dunbar, R. I. M. (2012). Latitudinal variation in light levels drives human visual system size. *Biology Letters* 8: 90–93.

– Pearce, E., Stringer, C., and Dunbar, R. I. M. (2013). New insights into differences in brain organisation between Neanderthals and anatomically modern humans. *Proceedings of the Royal Society, London* 280B.

– Reed, K. E. (1997). Early hominid evolution and ecological change through the African Plio-Pleistocene. *Journal of Human Evolution* 32: 289–322.

– Reed, K. E., and Russak, S. M. (2009). Tracking ecological change in relation to the emergence of *Homo* near the Plio-Pleistocene boundary. In: F. E. Grine, J. G. Fleagle and R. E. Leakey (eds) *The First Humans: Origin and Early Evolution of the Genus Homo*, pp. 159–71. Berlin: Springer.

– Reich, R., Green, R., Kircher, M. et al. (2010). Genetic history of an archaic hominin group from Denisova cave in Siberia. *Nature* 468: 1053–60.

– Rhodes, J. A., and Churchill, S. E. (2009). Throwing in the Middle and Upper Paleolithic: inferences from an analysis of humeral retroversion. *Journal of Human Evolution* 56: 1–10.

– Richards, M. P., and Trinkaus, E. (2009). Isotopic evidence for the diets of European Neanderthals and early modern humans. *Proceedings of the National Academy of Sciences, USA* 106: 16034–9.

– Richards, M. P., Pettitt, P. B., Trinkaus, E., Smith, F. H., Paunovi , M., and Karavani , I. (2000). Neanderthal diet at Vindija and Neanderthal predation: the evidence from stable isotopes. *Proceedings of the National Academy of Sciences, USA* 97: 7663–6.

– Richards, M. P., Jacobi R., Cook, J., Pettitt, P. B., and Stringer, C. B. (2005). Isotope evidence for the intensive use of marine foods by Late Upper Palaeolithic humans. *Journal of Human Evolution* 49: 390–94.

- Roach, N. T., Venkadesan, M., Rainbow, M. J., and Lieberman, D. E. (2013). Elastic energy storage in the shoulder and the evolution of high-speed throwing in *Homo*. *Nature* 498: 483–7.
- Roberts, M. B., Stringer, C. B., and Parfitt, S. A. (1994). A hominid tibia from Middle Pleistocene sediments at Boxgrove, UK. *Nature* 369: 311–13.
- Roberts, S. B. G., Dunbar, R. I. M., Pollet, T., and Kuppens, T. (2009). Exploring variations in active network size: constraints and ego characteristics. *Social Networks* 31: 138–46.
- Saladié, P., Huguet, R., Rodríguez-Hidalgo, A. et al. (2012). Intergroup cannibalism in the European Early Pleistocene: the range expansion and imbalance of power hypotheses. *Journal of Human Evolution* 63: 682–95.
- Schmitt, D., Churchill, S. E., and Hylander, W. L. (2003). Experimental evidence concerning spear use in Neandertals and early modern humans. *Journal of Archaeological Science* 30: 103–14.
- Sutcliffe, A., Dunbar, R. I. M., Binder, J., and Arrow, H. (2012). Relationships and the social brain: integrating psychological and evolutionary perspectives. *British Journal of Psychology* 103: 149–68.
- Thieme, H. (1998). The oldest spears in the world: Lower Palaeolithic hunting weapons from Schöningen, Germany. In: E. Carbonell, J. M. Bermudez de Castro, J. L. Arsuaga and X. P. Rodriguez (eds) *Los Primeros Pobladores de Europa* [*The First Europeans: Recent Discoveries and Current Debate*], pp. 169–93. Aldecoa: Burgos.
- Thieme, H. (2005). The Lower Palaeolithic art of hunting: the case of Schöningen 13 II–4, Lower Saxony, Germany. In: C. S. Gamble and M. Porr (eds) *The Hominid Individual in Context: Archaeological Investigations of Lower and Middle Palaeolithic Landscapes, Locales and Artefacts*, pp. 115–32. London: Routledge.
- Vallverdú, J., Allué, E., Bischoff, J. L. et al. (2005). Short human occupations in the Middle Palaeolithic level I of the Abric Romaní rock-shelter (Capellades, Barcelona, Spain). *Journal of Human Evolution* 48: 157–74.
- Vaquero, M., and Pastó, I. (2001). The definition of spatial units in Middle Palaeolithic sites: the hearth related assemblages. *Journal of Archaeological Science* 28: 1209–20.
- Vaquero, M., Vallverdú, J., Rosell, J., Pastó, I., and Allué, E. (2001). Neandertal behavior at the Middle Palaeolithic site of Abric Romani, Capellades, Spain. *Journal of Field Archaeology* 28: 93–114.
- Weaver, T. D., and Hublin, J.-J. (2009). Neandertal birth canal shape and the evolution of human childbirth. *Proceedings of the National Academy of Sciences, USA* 106: 8151–6.

- Wilkins, J., Schoville, B. J., Brown, K. S., and Chazan, M. (2012). Evidence for early hafted hunting technology. *Science* 338: 942–6.
- Zollikofer, C. P. E., Ponce de León, M. S., Vandermeersch, B., and Lévêque, F. (2002). Evidence for interpersonal violence in the St Césaire Neanderthal. *Proceedings of the National Academy of Sciences, USA* 99: 6444–8.

CHAPTER 7: THE FOURTH TRANSITION: MODERN HUMANS

- Aiello, L. C. (1996). Terrestriality, bipedalism, and the origin of language. In: G. Runciman, J. Maynard-Smith and R. I. M. Dunbar (eds) *Evolution of Social Behaviour Patterns in Primates and Man*, pp. 269–89. Oxford: Oxford University Press.
- Aiello, L. C., and Dunbar, R. I. M. (1993). Neocortex size, group size and the evolution of language. *Current Anthropology* 34: 184–93.
- Aiello, L. C., and Wheeler, P. (2003). Neanderthal thermoregulation and the glacial climate. In: T. H. van Andel and W. Davies (eds.) *Neanderthals and Modern Humans in the European Landscape During the Late Glaciation*. Cambridge: Cambridge University Press.
- Arensburg, B., Tillier, A. M., Vandermeersch, B., Duday, H., Schepartz, L. A., and Rak, Y. (1989). A Middle Palaeolithic human hyoid bone. *Nature* 338, 758–60.
- Atkinson, Q. D., Gray, R. D., and Drummond, A. J. (2009). Bayesian coalescent inference of major human mitochondrial DNA haplogroup expansions in Africa. *Proceedings of the Royal Society, London* 276B: 367–73.
- Bailey, D., and Geary, D. (2009). Hominid Brain Evolution. *Human Nature* 20: 67–79.
- Balzeau, A., Holloway, R. L., and Grimaud-Hervé, D. (2012). Variations and asymmetries in regional brain surface in the genus *Homo*. *Journal of Human Evolution* 62: 696–706.
- Barton, R. A., and Venditti, C. (2013). Human frontal lobes are not relatively large. *Proceedings of the National Academy of Sciences, USA* 111: 942–47.
- Bruner, E., Manzi, G., and Arsuaga, J. L. (2003). Encephalization and allometric trajectories in the genus *Homo*: Evidence from the Neandertal and modern lineages. *Proceedings of the National Academy of Sciences, USA* 100: 15335–40.
- Burke, A. (2012). Spatial abilities, cognition and the pattern of Neanderthal and modern human dispersals. *Quaternary International* 247: 230–35.
- Caspari, R., and Lee, S.-H. (2004). Older age becomes common late in

human evolution. *Proceedings of the National Academy of Sciences, USA* 101: 10895–900.

– Comas, I., Coscolla, M., Luo, T. et al. (2013). Out-of-Africa migration and Neolithic coexpansion of *Mycobacterium tuberculosis* with modern humans. *Nature Genetics* 45: 1176–82.

– Cowlishaw, C., and : Dunbar, R. I. M. (2000). *Primate Conservation Biology*. Chicago IL: Chicago University Press.

– DaGusta, D., Gilbert, W. H., and Turner, S. P. (1999). Hypoglossal canal size and hominid speech. *Proceedings of the National Academy of Sciences, USA* 96: 1800–804.

– Deacon, T. W. (1995). *The Symbolic Species: The Coevolution of Language and the Human Brain*. Harmondsworth: Allen Lane.

– Dean, C., Leakey, M. G., Reid, D. et al. (2001). Growth processes in teeth distinguish modern humans from *Homo erectus* and earlier hominins. *Nature* 414: 628–31.

– Dobson, S. D. (2009). Socioecological correlates of facial mobility in nonhuman anthropoids. *American Journal of Physical Anthropology* 139: 413–20.

– Dobson, S. D. (2012). Face to face with the social brain. *Philosophical Transactions of the Royal Society* 367B: 1901–8.

– Dobson, S. D., and Sherwood, C. C. (2011). Correlated evolution of brain regions involved in producing and processing facial expressions in anthropoid primates. *Biology Letters* 7: 86–8.

– Dunbar, R. I. M. (2012). Bridging the bonding gap: the transition from primates to humans. *Philosophical Transactions of the Royal Society, London* 367B: 1837–46.

– Dunbar, R. I. M., and Shi, J. (2013). Time as a constraint on the distribution of feral goats at high latitudes. *Oikos* 122: 403–10.

– Dunsworth, H. M., Warrener, A. G., Deacon, T., Ellison, P. T., and Pontzer, H. (2012). Metabolic hypothesis for human altriciality. *Proceedings of the National Academy of Sciences, USA* 109: 15212–16.

– Enard, W., Przeworski, M., Fisher, S. E. et al. (2002). Molecular evolution of FOXP2, a gene involved in speech and language. *Nature* 418: 869–72.

– Féblot-Augustins, J. (1993). Mobility strategies in the Late Middle Palaeolithic of central Europe and western Europe: elements of stability and variability. *Journal of Anthropological Archaeology* 12: 211–65.

– Finlay, B. L., Darlington, R. B., and Nicastro, N. (2001). Developmental structure in brain evolution. *Behavioral and Brain Sciences* 24: 263–308.

– Finlayson, C. (2010). *The Humans Who Went Extinct: Why Neanderthals Died Out and We Survived*. Oxford: Oxford University Press.

– Fisher, S. E., and Marcus, G. F. (2006). The eloquent ape: genes, brains and the evolution of language. *Nature Reviews Genetics* 7: 9–20.

– Freeberg, T. M. (2006). Social complexity can drive vocal complexity. *Psychological Science* 17: 557–61.

– Goebel, T., Waters, M. R., and O'Rourke, D. H. (2008). The late Pleistocene dispersal of modern humans in the Americas. *Science* 319: 1497–502.

– Haesler, S., Rochefort, C., Georgi, B., Licznerski, P., Osten, P., and Scharff, C. (2007). Incomplete and inaccurate vocal imitation after knockdown of FoxP2 in songbird basal ganglia nucleus Area X. *PLoS Biology* 5: e321.

– Helgason, A., Hickey, E., Goodacre, S. et al. (2001). mtDNA and the islands of the North Atlantic: estimating the proportions of Norse and Gaelic ancestry. *American Journal of Human Genetics* 68: 723–37.

– Helgason, A., Sigurðardóttir, S., Gulcher, J. R., Ward, R., and Stefánsson, K. (2000). mtDNA and the origin of the Icelanders: deciphering signals of recent population history. *American Journal of Human Genetics* 66: 999–1016.

– Henn, B., Gignoux, C., Jobin, M. et al. (2011). Hunter-gatherer genomic diversity suggests a southern African origin for modern humans. *Proceedings of the National Academy of Sciences, USA* 108: 5154–62.

– Horan, R. D., Bulte, E., and Shogren, J. F. (2005). How trade saved humanity from biological exclusion: an economic theory of Neanderthal extinction. *Journal of Economic Behavior and Organization* 58: 1–29.

– Ingman, M., Kaessmann, H., Pääbo, S., and Gyllensten, U. (2000). Mitochondrial genome variation and the origin of modern humans. *Nature, London* 408: 708–13.

– Joffe, T. H. (1997). Social pressures have selected for an extended juvenile period in primates. *Journal of Human Evolution* 32: 593–605.

– Jungers, W. L., Pokempner, A., Kay, R. F., and Cartmill, M. (2003). Hypoglossal canal size in living hominoids and the evolution of human speech. *Human Biology* 75: 473–84.

– Kay, R. F., Cartmill, M., and Balow, M. (1998). The hypoglossal canal and the origin of human vocal behaviour. *Proceedings of the National Academy of Sciences, USA* 95: 5417–19.

– Klein, R. G. (2000). *The Human Career: Human Biological and Cultural Origins*, 3rd edition. Chicago: Chicago University Press.

– Krause, J., Lalueza-Fox, C., Orlando, L. et al. (2007). The derived *FoxP2* variant of modern humans was shared with Neanderthals. *Current Biology* 17: 1908–12.

– Lahr, M. M., and Foley, R. (1994). Multiple dispersals and modern human origins. *Evolutionary Anthropology* 3: 48–60.

– Lewis, P. A., Rezaie, R., Browne, R., Roberts, N., and Dunbar, R. I. M. (2011).

Ventromedial prefrontal volume predicts understanding of others and social network size. *NeuroImage* 57: 1624–9.

- McComb, K., and Semple, S. (2005). Coevolution of vocal communication and sociality in primates. *Biology Letters* 1: 381–5.

- MacLarnon, A., and Hewitt, G. (1999). The evolution of human speech: the role of enhanced breathing control. *American Journal of Physical Anthropology* 109: 341–63.

- Martín-González, J., Mateos, A., Goikoetxea, I., Leonard, W., and Rodríguez, J. (2012). Differences between Neandertal and modern human infant and child growth models. *Journal of Human Evolution* 63: 140–49.

- Martinez, I., Rosa, M., Jarabo, P. et al. (2004). Auditory capacities in Middle Pleistocene humans from the Sierra de Atapuerca in Spain. *Proceedings of the National Academy of Sciences, USA* 101: 9976–81.

- Noble, W., and Davidson, I. (1991). The evolutionary emergence of modern human behaviour. I. Language and its archaeology. *Man* 26: 222–53.

- Osaka City University (2011). Catalogue of Fossil Hominids Database. http://gbs.ur-plaza.osaka-cu.ac.jp/kaseki/index.html

- Powell, J., Lewis, P. A., Dunbar, R. I. M., García-Fiñana, M., and Roberts, N. (2010). Orbital prefrontal cortex volume correlates with social cognitive competence. *Neuropsychologia* 48: 3554–62.

- Richards, M. P., and Trinkaus, E. (2009). Isotopic evidence for the diets of European Neanderthals and early modern humans. *Proceedings of the National Academy of Sciences, USA* 38: 16034–9.

- Semendeferi, K., Damasio, H., Frank, R., and Van Hoesen, G. W. (1997). The evolution of the frontal lobes: a volumetric analysis based on three-dimensional reconstructions of magnetic resonance scans of human and ape brains. *Journal of Human Evolution* 32: 375–88.

- Slimak, L., and Giraud, Y. (2007). Circulations sur plusieurs centaines de kilomètres durant le Paléolithique moyen. Contribution à la connaissance des sociétés néandertaliennes. *Comptes Rendus Palevol* 6: 359–68.

- Smith, T. M., Tafforeau, P., Reid, D. J. et al. (2007). Earliest evidence of modern human life history in North African early *Homo sapiens*. *Proceedings of the National Academy of Sciences, USA* 104: 6128–33.

- Stedman, H. H., Kozyak, B. W., Nelson, A. et al. (2004). Myosin gene mutation correlates with anatomical changes in the human lineage. *Nature* 428: 415–18.

- Stoneking, M. (1993). DNA and recent human evolution. *Evolutionary Anthropology* 2: 60–73.

- Shultz, A., and Maslin, M. (2013). Early human speciation, brain expansion and dispersal influenced by African climate pulses. *PLoS One* 8: e76750.

- Thomas, M. G., Stumpf, M. P. H., and Härke, H. (2006). Evidence for an apartheid-like social structure in early Anglo-Saxon England. *Proceedings of the Royal Society, London* 273B: 2651–7.
- Toups, M. A., Kitchen, A., Light, J. E., and Reed, D. L. (2011). Origin of clothing lice indicates early clothing use by anatomically modern humans in Africa. *Molecular Biology and Evolution* 28: 29–32.
- Uomini, N. T. (2009). The prehistory of handedness: archaeological data and comparative ethology. *Journal of Human Evolution* 57: 411–19.
- Uomini, N., and Meyer, G. (2013). Shared brain lateralization patterns in language and Acheulean stone tool production: a functional transcranial Doppler ultrasound study. *PLoS One* 8: e72693.
- Williams, A., and Dunbar, R. I. M. (2013). Big brains, meat, tuberculosis, and the nicotinamide switches: co-evolutionary relationships with modern repercussions? *International Journal of Tryptophan Research* 3: 73–88.
- Zerjal, T., Xue, Y., Bertorelle, G. et al. (2003). The genetic legacy of the Mongols. *American Journal of Human Genetics* 72: 717–21.

CHAPTER 8: HOW KINSHIP, LANGUAGE AND CULTURE CAME TO BE

- Bader, N. O., and Lavrushin Y. A. (eds) (1998). *Upper Palaeolithic Site Sungir (graves and environment)* (*Posdnepaleolitischeskoje posselenije Sungir*). Moscow: Scientific World.
- Bailey, D., and Geary, D. (2009). Hominid Brain Evolution. *Human Nature* 20: 67–9.
- Boyer, P. (2001). *Religion Explained: The Human Instincts That Fashion Gods, Spirits and Ancestors*. London: Weidenfeld & Nicholson.
- Burton-Chellew, M., and Dunbar, R. I. M. (2011). Are affines treated as biological kin? A test of Hughes' hypothesis. *Current Anthropology* 52: 741–6.
- Cashmore, L., Uomini, N., and Chapelain, A. (2008). The evolution of handedness in humans and great apes: a review and current issues. *Journal of Anthropological Science* 86: 7–35.
- Conard, N. J. (2003). Palaeolithic ivory sculptures from southwestern Germany and the origins of figurative art. *Nature* 426: 830–83.
- Curry, O., and Dunbar, R. I. M. (2013). Do birds of a feather flock together? The relationship between similarity and altruism in social networks. *Human Nature* 24: 336–47.
- Curry, O., Roberts, S., and Dunbar, R. I. M. (2013). Altruism in social networks: evidence for a 'kinship premium'. *British Journal of Psychology* 104: 283–95.

– D'Errico, F., Henshilwood, C., Vanhaeren, M., and van Niekerk, K. (2005). *Nassarius kraussianus* shell beads from Blombos Cave: evidence for symbolic behaviour in the Middle Stone Age. *Journal of Human Evolution* 48: 3–24.

– Deacon, T. W. (1995). *The Symbolic Species: The Coevolution of Language and the Human Brain*. Harmondsworth: Allen Lane.

– Dunbar, R. I. M. (1993). Coevolution of neocortex size, group size, and language in humans. *Behavioral Brain Sciences* 16: 681–735.

– Dunbar, R. I. M. (1995). On the evolution of language and kinship. In: J. Steele and S. Shennan (eds) *The Archaeology of Human Ancestry: Power, Sex and Tradition*, pp. 380–96. London: Routledge.

– Dunbar, R. I. M. (1996). *Grooming, Gossip and the Evolution of Language*. London: Faber & Faber.

– Dunbar, R. I. M. (2008). Mind the gap: or why humans aren't just great apes. *Proceedings of the British Academy* 154: 403–23.

– Dunbar, R. I. M. (2009). Why only humans have language. In: R. Botha and C. Knight (eds) *The Prehistory of Language*, pp. 12–35. Oxford: Oxford University Press.

– Dunbar, R. I. M. (2013). The origin of religion as a small scale phenomenon. In: S. Clark and R. Powell (eds) *Religion, Intolerance and Conflict: A Scientific and Conceptual Investigation*, pp. 48–66. Oxford: Oxford University Press.

– Fincher, C. L., and Thornhill, R. (2008). Assortative sociality, limited dispersal, infectious disease and the genesis of the global pattern of religion diversity. *Proceedings of the Royal Society, London* 275B: 2587–94.

– Fincher, C. L., Thornhill, R., Murray, D. R., and Schaller, M. (2008). Pathogen prevalence predicts human cross-cultural variability in individualism/collectivism. *Proceedings of the Royal Society, London* 275B: 1279–85.

– Frankel, B. G., and Hewitt, W. E. (1994). Religion and well-being among Canadian university students: the role of faith groups on campus. *Journal of the Scientific Study of Religion* 33: 62–73.

– Hamilton, W. D. (1964). The genetical evolution of social behaviour. I, II. *Journal of Theoretical Biology* 7: 1–52.

– Henshilwood, C. S., d'Errico, F., van Niekerk, K. L. et al. (2011). A 100,000-Year-Old Ochre-Processing Workshop at Blombos Cave, South Africa. *Science* 334: 219–22.

– Henshilwood, C. S., d'Errico, F., Yates, R. et al. (2002). Emergence of modern human behavior: Middle Stone Age engravings from South Africa. *Science* 295: 1278–80.

– Hughes, A. (1988). *Kinship and Human Evolution*. Oxford: Oxford University Press.

– Klein, R. G. (2000). *The Human Career: Human Biological and Cultural Origins*, 3rd edition. Chicago: Chicago University Press.
– Koenig, H. G., and Cohen, H. J. (eds) (2002). *The Link Between Religion and Health: Psychoneuroimmunology and the Faith Factor*. Oxford University Press: Oxford.
– Kudo, H., and Dunbar, R. I. M. (2001). Neocortex size and social network size in primates. *Animal Behaviour* 62: 711–22.
– Layton, R., O'Hara, S., and Bilsborough, A. (2012). Antiquity and social functions of multilevel social organization among human hunter-gatherers. *International Journal of Primatology* 33: 1215–45.
– Lehmann, J., Lee, P. C., and Dunbar, R. I. M. (2013). Unravelling the evolutionary function of communities. In: R. I. M. Dunbar, C. Gamble and J. A. J. Gowlett (eds) *Lucy to Language: The Benchmark Papers*, pp. 245–76. Oxford: Oxford University Press.
– Lewis-Williams, D. (2002). *The Mind in the Cave*. London: Thames & Hudson.
– Mesoudi, A., Whiten, A., and Dunbar, R. I. M. (2006). A bias for social information in human cultural transmission. *British Journal of Psychology* 97: 405–23.
– Mickes, L., Darby, R. S., Hwe, V. et al. (2013). Major memory for microblogs. *Memory and Cognition* 41: 481–89.
– Miller, G. (1999). Sexual selection for cultural displays. In R. I. M. Dunbar, C. Knight and C. Power (eds) *The Evolution of Culture*, pp. 71–91. Edinburgh: Edinburgh University Press.
– Nettle, D. (1999). *Linguistic Diversity*. Oxford: Oxford University Press.
– Nettle, D., and Dunbar, R. I. M. (1997). Social markers and the evolution of reciprocal exchange. *Current Anthropology* 38: 93–9.481–89.
– Osaka City University (2011). Catalogue of Fossil Hominids Database. http://gbs.ur-plaza.osaka-cu.ac.jp/kaseki/index.html
– Palmer, C. T. (1991). Kin selection, reciprocal altruism and information sharing among Maine lobstermen. *Ethology and Sociobiology* 12: 221–35.
– Redhead, G., and Dunbar, R. I. M. (2013). The functions of language: an experimental study. *Evolutionary Psychology* 11: 845–54.
– Rouget, G. (1985). *Music and Trance: A Theory of the Relations Between Music and Possession*. Chicago: University of Chicago Press.
– Silk, J. B. (1980). Adoption and kinship in Oceania. *American Anthropologist* 82: 799–820.
– Silk, J. B. (1990). Which humans adopt adaptively and why does it matter? *Ethology and Sociobiology* 11: 425–6.
– Thornhill, R., Fincher, C. L., and Aran, D. (2009). Parasites, democratization,

and the liberalization of values across contemporary countries. *Biology Reviews* 84: 113–31.

‒ Wiessner, P. (2002). Hunting, healing, and *hxaro* exchange: a long-term perspective on !Kung (Ju/'hoansi) large-game hunting. *Evolution and Human Behavior* 23: 1–30.

CHAPTER 9: THE FIFTH TRANSITION: THE NEOLITHIC AND BEYOND

‒ Andelman, S. (1986). Ecological and social determinants of cercopithecine mating patterns. In: D. I. Rubenstein and R. W. Wrangham (eds) *Ecological Aspects of Social Evolution*, pp. 201–16. Princeton NJ: Princeton University Press.

‒ Atkinson, Q. D., and Bourrat, P. (2011). Beliefs about God, the afterlife and morality support the role of supernatural policing in human cooperation. *Evolution and Human Behavior* 32: 41–9.

‒ Bourrat, P., Atkinson, Q. D., and Dunbar, R. I. M. (2011). Supernatural punishment and individual social compliance across cultures. *Religion, Brain and Behavior* 1: 119–34.

‒ Bowles, S. (2009). Did warfare among ancestral hunter-gatherers affect the evolution of human social behaviors? *Science* 324: 1293–8.

‒ Bowles, S. (2011). Cultivation of cereals by the first farmers was not more productive than foraging. *Proceedings of the National Academy of Sciences, USA* 108: 4760–65.

‒ Bugos, P., and McCarthy, L. (1984). Ayoreo infanticide: a case study. In: G. Hausfater and S. B. Hrdy (eds) *Infanticide: Comparative and Evolutionary Perspectives*, pp. 503–20. Hawthorne: Aldine de Gruyter.

‒ Caspari, R., and Lee, S.-H. (2004). Older age becomes common late in human evolution. *Proceedings of the National Academy of Sciences, USA* 101: 10895–900.

‒ Cohen, M. N., and Crane-Kramer, G. (2007). *Ancient Health: Skeletal Indicators of Agricultural and Economic Intensification*. Gainesville, FL: University Press of Florida.

‒ Coward, F., and Dunbar, R. I. M. (2013). Communities on the edge of civilisation. In: R. I. M. Dunbar, C. Gamble and J. A. J. Gowlett (eds.) *Lucy to Language: The Benchmark Papers*, pp. 380–405. Oxford: Oxford University Press.

‒ Curry, O., and Dunbar, R. I. M. (2011). Altruism in networks: the effect of connections. *Biology Letters* 7: 651–3.

- Curry, O., and Dunbar, R. I. M. (2013). Do birds of a feather flock together? The relationship between similarity and altruism in social networks. *Human Nature* 24: 336–47.

- Curry, O., and Dunbar, R. I. M. (2013). Sharing a joke: the effects of a similar sense of humor on affiliation and altruism. *Evolution and Human Behavior* 34: 125–9.

- Daly, M., and Wilson, M. (1981). Abuse and neglect of children in evolutionary perspective. In: R. D. Alexander and D. W. Tinkle (eds) *Natural Selection and Social Behavior*, pp. 405–16. New York: Chiron Press.

- Daly, M., and Wilson, M. (1984). A sociobiological analysis of human infanticide. In: G. Hausfater and S. B. Hrdy (eds) *Infanticide: Comparative and Evolutionary Perspectives*, pp. 487–502. New York: Aldine de Gruyter.

- Daly, M., and Wilson, M. (1985). Child abuse and other risks of not living with both parents. *Ethology and Sociobiology* 6: 197–210.

- Daly, M., and Wilson, M. (1988). Evolutionary psychology and family homicide. *Science* 242: 519–24.

- Diamond, J. (2002). Evolution, consequences and future of plant and animal domestication. *Nature* 418: 700–707.

- Dietrich, O., Heun, M., Notroff, J., Schmidt, K., and Zarnkow, M. (2012). The role of cult and feasting in the emergence of Neolithic communities. New evidence from Göbekli Tepe, south-eastern Turkey. *Antiquity* 86: 674–95.

- Dunbar, R. I. M. (2000). Male mating strategies: a modelling approach. In: P. Kappeler (ed.) *Primate Males*, pp. 259–68. Cambridge: Cambridge University Press.

- Dunbar, R. I. M. (2010). Deacon's dilemma: the problem of pairbonding in human evolution. In: R. I. M. Dunbar, C. Gamble and J. A. J. Gowlett (eds.) *Social Brain, Distributed Mind*, pp. 159–79. Oxford: Oxford University Press.

- Dunbar, R. I. M. (2012). *The Science of Love and Betrayal*. London: Faber & Faber.

- Dunbar, R. I. M. (2012). Social cognition on the internet: testing constraints on social network size. *Philosophical Transactions of the Royal Society, London* 367B: 2192–2201.

- Dunbar, R. I. M. (2013). The origin of religion as a small scale phenomenon. In: S. Clark and R. Powell (eds) *Religion, Intolerance and Conflict: A Scientific and Conceptual Investigation*, pp. 48–66. Oxford: Oxford University Press.

- Dunbar, R. I. M., Lehmann, J., Korstjens, A. H., and Gowlett, J. A. J. (2014). The road to modern humans: time budgets, fission-fusion sociality, kinship and the division of labour in hominin evolution. In: R. I. M. Dunbar, C. Gamble and J. A. J. Gowlett (eds) *Lucy to Language: The Benchmark Papers*, pp. 333–55. Oxford: Oxford University Press.

– Ember, C. R., Adem, T. A., and Skoggard, I. (2013). Risk, uncertainty, and violence in Eastern Africa: a regional comparison. *Human Nature* 24: 33–58.

– Fehr, E., and Gächter, S. (2002). Altruistic punishment in humans. *Nature* 415: 137–40.

– Fibiger, L., Ahlström, T., Bennike, P., and Schulting, R. J. (2013). Patterns of violence-related skull trauma in Neolithic southern Scandinavia. *American Journal of Physical Anthropology* 150: 190–202.

– Fisher, H. E., Aron, A., and Brown, L. L. (2006). Romantic love: a mammalian brain system for mate choice. *Philosophical Transactions of the Royal Society, London* 361B: 2173–86.

– Harcourt, A. H., Harvey, P. H., Larson, S. G., and Short, R. V. (1981). Testis weight, body weight and breeding system in primates. *Nature* 293: 55–7.

– Henrich, J., Ensminger, J., McElreath, R. et al. (2010). Markets, religion, community size, and the evolution of fairness and punishment. *Science* 327: 1480–84.

– Hewlett, B. S. (1988). Sexual selection and paternal investment among Aka pygmies. In: L. Betzig, M. Borgerhoff-Mulder and P. Turke (eds) *Human Reproductive Behaviour*, pp. 263–75. Cambridge: Cambridge University Press.

– Jankowiak, W. R., and Fischer, E. F. (1992). A cross-cultural perspective on romantic love. *Ethnology* 31: 149–55.

– Johnson, A. W., and Earle, T. K. (2001). *The Evolution of Human Societies: From Foraging Group to Agrarian State*, 2nd edition. Palo Alto, CA: Stanford University Press.

– Johnson, D. D. P. (2005). God's punishment and public goods: a test of the supernatural punishment hypothesis in 186 world cultures. *Human Nature* 16: 410–46.

– Johnson, D. D. P., and Bering, J. (2009). Hand of God, mind of man. In: J. Schloss and M. J. Murray (eds), *The Believing Primate: Scientific, Philosophical, and Theological Reflections on the Origin of Religion*, pp. 26–44. Oxford: Oxford University Press.

– Knott, C. D., and Kahlenberg, S. M. (2007). Orangutans in perspective: forced copulations and female mating resistance. In: C. J. Campbell, A. Fuentes, K. C. MacKinnon, M. Panger and S. K. Bearder (2007). *Primates in Perspective*, pp. 290–305. New York: Oxford University Press.

– Lehmann, J., Korstjens, A. H., and Dunbar, R. I. M. (2007). Fission–fusion social systems as a strategy for coping with ecological constraints: a primate case. *Evolutionary Ecology* 21: 613–34.

– Lukas, D., and Clutton-Brock, T. H. (2013). The evolution of social monogamy in mammals. *Science* 341: 526–30.

- Manning, J. T. (2002). *Digit Ratio: A Pointer to Fertility, Health, and Behavior*. New Brunswick, NJ: Rutgers University Press.
- Mesnick, S. L. (1997). Sexual alliances: evidence and evolutionary implications. In: P. A. Gowaty (ed.) *Feminism and Evolutionary Biology*, pp. 207–60. London: Chapman & Hall.
- Munro, N. D., and Grosman L. (2010). Early evidence (ca. 12,000 B.P.) for feasting at a burial cave in Israel. *Proceedings of the National Academy of Sciences, USA* 107: 15362–6.
- Naroll, R. (1956). A preliminary index of social development. *American Anthropologist* 58: 687–715.
- Nelson, E., Rolian, C., Cashmore, L., and Shultz, S. (2011). Digit ratios predict polygyny in early apes, *Ardipithecus*, Neanderthals and early modern humans but not in Australopithecus. *Proceedings of the Royal Society, London* 278B: 1556–63.
- Nettle, D., and Dunbar, R. I. M. (1997). Social markers and the evolution of reciprocal exchange. *Current Anthropology* 38: 93–9.
- Norenzayan, A., and Shariff, A. F. (2008). The origin and evolution of religious prosociality. *Science* 322: 58–62.
- Opie, C., Atkinson, Q. D., Dunbar, R. I. M., and Shultz, S. (2013). Male infanticide leads to social monogamy in primates. *Proceedings of the National Academy of Sciences, USA* 110: 13328–32.
- Palchykov, V., Kaski, K., Kertész, J., Barabási, A.-L., and Dunbar, R. I. M. (2012). Sex differences in intimate relationships. *Scientific Reports* 2: 320.
- Palchykov, V., Kertész, J., Dunbar, R. I. M., and Kaski, K. (2013). Close relationships: a study of mobile communication records. *Journal of Statistical Physics* 151: 735–44.
- Pérusse, D. (1993). Cultural and reproductive success in industrial societies: testing the relationship at the proximate and ultimate levels. *Behavioral and Brain Sciences* 16: 267–322.
- Putnam, R. D. (2000). *Bowling Alone: The Collapse and Revival of American Community*. New York: Simon and Schuster.
- Reno, P. L., Meindl, R. S, McCollum, M. A., and Lovejoy, C. O. (2003). Sexual dimorphism in *Australopithecus afarensis* was similar to that of modern humans. *Proceedings of the National Academy of Sciences, USA* 100: 9404–9
- Roberts, S. B. G., and Dunbar, R. I. M. (2011). The costs of family and friends: an 18-month longitudinal study of relationship maintenance and decay. *Evolution and Human Behavior* 32: 186–97.
- Roberts, S. B. G., and Dunbar, R. I. M. (2011). Communication in social networks: effects of kinship, network size and emotional closeness. *Personal Relationships* 18: 439–52.

– Roes, F. L., and Raymond, M. (2003). Belief in moralizing gods. *Evolution and Human Behavior* 24: 126–35.
– van Schaik, C. P., and Dunbar, R. I. M. (1991). The evolution of monogamy in large primates: a new hypothesis and some critical tests. *Behaviour* 115: 30–62.
– Sosis, R., and Alcorta, C. (2003). Signaling, solidarity, and the sacred: the evolution of religious behavior. *Evolutionary Anthropology* 12: 264–74.
– Stanford, D., and Bradley, B. (2002). Ocean trails and prairie paths? Thoughts about Clovis origins. In: N. G. Jablonski (ed.) *The First Americans: The Pleistocene Colonization of the New World*, pp. 255–71. San Francisco: California Academy of Sciences.
– Sutcliffe, A., Dunbar, R. I. M., Binder, J., and Arrow, H. (2012). Relationships and the social brain: integrating psychological and evolutionary perspectives. *British Journal of Psychology* 103: 149–68.
– Ulijaszek, S. J. (1991). Human dietary change. *Philosophical Transactions of the Royal Society, London* 334B: 271–9.
– Voland, E., and Engel, C. (1989). Women's reproduction and longevity in a premodern population (Ostfriesland, Germany, 18th century). In: A. E. Rasa, C. Vogel and E. Voland (eds.) *The Sociobiology of Sexual and Reproductive Strategies*, pp. 194–205. London: Chapman & Hall.
– Walker, R. S., and Bailey, D. H. (2013). Body counts in lowland South American violence. *Evolution and Human Behavior* 34: 29–34.
– Watts, D. P. (1989). Infanticide in mountain gorillas: new cases and a reconsideration of the evidence. *Ethology* 81: 1–18.
– Wilson, M., and Mesnick, S. L. (1997). An empirical test of the bodyguard hypothesis. In: P. A. Gowaty (ed.) *Feminism and Evolutionary Biology*. London: Chapman & Hall.

Notes

1. In apes, the thigh bone hangs straight down from its joint with the pelvis; as a result, when apes walk bipedally, they have to swing the body over each knee to maintain balance, giving a waddling style of walking. In bipedal hominins, the thigh bones angle in to give a knock-kneed appearance that allows the body's centre of gravity to be above the knee no matter which leg is in contact with the ground.

2. Mitochondria are the tiny powerhouses inside conventional cells that provide the energy that keeps the cell working. They have their own DNA, which is different from the more conventional nuclear DNA that creates our bodies. Because the mitochondria are part of the cell cytoplasm that surrounds the cell nucleus (where the conventional DNA that determines who we are is housed), they are inherited only down the female line – sperm has no cytoplasm and simply consists of a cell nucleus. Mitochondria were originally free-living bacteria that found their way into the cells of multicellular organisms very early on during the evolution of life on earth, where they then established a remarkably successful symbiotic relationship that allowed them to survive and reproduce very effectively courtesy of the host cells.

3. Humans didn't reach the Americas until much later (around 16,000 years ago), and when they did they probably either walked there across the exposed Bering Strait, and then spread very rapidly through North and then South America, possibly in three successive waves (the Eskimos being the third and last of these). Recently, a controversial claim has been made that there was an earlier invasion of North America from Europe by Solutrean people from southwest Europe who crossed the Atlantic by hugging the polar pack ice perhaps as early as 20,000 years ago to become

the enigmatic Clovis Folk of the south and east of the USA (the Solutrean Atlantic Hypothesis: Stanford and Bradley 2002). The presumption is that the Clovis Folk went extinct at the hands of the later Siberian invaders, or were wholly absorbed by them (to the point that they left no genetic trace). Inevitably, perhaps, the Solutrean hypothesis has been the subject of vigorous debate.

4. This oceanic trench is biologically so significant that, aside from a few birds that occur on both sides, it divides two major faunal zones, the Australian and the Asian (including the rest of the world). It is known as the Wallace Line after Alfred Russel Wallace, the co-discoverer of the modern theory of evolution.

5. Although anthropology technically means 'the study of mankind', anthropologists mostly consider themselves to be students of culture and cultural variation across human societies. The term *panthropology* trades on this meaning to refer to the study of culture in apes.

6. By *para-fictional*, I mean a world that is genuinely fictional, but which we might nonetheless firmly believe exists.

7. These data for this figure, and all subsequent graphs based on this figure, use the estimates published by De Miguel and Heneberg (2001). Palaeontologists will always disagree as to whose dataset is the best and most accurate, but in reality the differences are trivial, and so we will simply rely on this particular dataset and leave it at that.

8. 'Analogue' because they are based on the assumption that the living species provides a reasonable analogy to the fossil one by virtue of some appropriate ecological or taxonomic similarity.

9. Plantar refers to having the whole of the sole of the foot in contact with the ground when walking. The human foot, with its toes half-hidden within the foot, is unique among the primates, all of whom have hand-like feet designed to grasp the branches of trees. The human plantar foot is designed specifically to allow our striding form of locomotion.

CHAPTER 2: THE BASES OF PRIMATE SOCIALITY

1. Grueter et al. (2013) recently suggested that grooming time is determined by terrestriality in primates, and not by group size, and hence reflects the demands of hygiene and not sociality. Unfortunately, their analysis is based on a number of rather odd misconceptions that need not detain us; the

more important point is that their own data unequivocally confirm, for Old World monkeys and apes, the relationship shown in Fig. 2.1 independently of terrestriality (see Dunbar and Lehmann 2013). That grooming helps keep the fur clean and free of debris and parasites is not in dispute; the issue is whether, in addition to this, grooming is used to service social relationships.

2. This is done by reconstructing the ancestral states for the variables of interest right through the primate phylogeny – the family tree based on genetic relatedness that links all the living species back to their common ancestor some 65 or more million years ago – and then investigating the order in which switches from one state to another occurred along the various evolutionary pathways leading to modern species. To be the cause of a switch in behaviour or social organization, a change in selection pressure has to precede the change in the behaviour it influences. The statistical methods used to do this are based on Bayesian statistics and represent a major new advance over how these kinds of problems have been approached in the past. Further details about the new Bayesian methods can be found in Huelsenbeck et al. (2001).

3. This is a consequence of the fact that it is the baby sucking on the breast that suppresses the menstrual hormone system. Once the baby is being weaned and its rate of sucking drops below a specific rate, the menstrual hormonal system kicks back into action. This is why the female returns to menstrual cycling immediately if the baby dies or is killed, even in humans. All mammals have this mechanism.

4. Adult male gorillas are more than twice the size of adult females.

CHAPTER 3: THE ESSENTIAL FRAMEWORK

1. Lifehistory refers to the demographic variables (fertility, duration of gestation and lactation, age at puberty, and longevity) that determine an individual's ability to contribute offspring to the next generation, and through this to influence population dynamics as well as evolution.

2. The cerebellum is the small brain-like bulb that sits just below the back of the brain. It is a very ancient part of the brain whose main function seems to be to coordinate cognitive processing in different parts of the brain, so that a sequence of steps is executed in the right order at the right time. For

this reason, it is heavily involved in locomotion and ensures that our limb movements are coordinated and in balance. The cerebellum is relatively larger in the human lineage than in other primates, in part because of the complexity of bipedal locomotion. However, its functions are not just limited to locomotion, but also seem to involve the coordination of thought processes more generally.

3. This study (Hamilton et al. 2007) obtained a scaling ratio of 3.8, and concluded that the true scaling ratio was 4. In fact, this turns out to be because they anchored the sequence of groupings at the individual (in effect, a group size of 1) rather than at the first level of true grouping (the 5 level). If we ignore the individual, their own data yield a scaling ratio of 3.3 – exactly as in our analysis (see Zhou et al. 2005) – with a starting point of exactly 5 and layer sizes that match exactly the sequence we found in our data (see Hill et al. 2008; Lehmann et al. 2014).

4. Leaf cells are protected by celluloses that mammals cannot easily digest. Folivorous primates, like ruminant ungulates, rely on fermentation by bacteria to break down the cell wall and release the nutrients. The fermentation process takes a lot of time, during which the animal has to remain inactive because being active suppresses bacterial fermentation.

5. A genus is a collection of closely related species who typically share the same dietary and body plan.

6. Full details of this model can be found in Lehmann et al. (2008b). For the benefit of those who would like to know, the equations are:

% Feed time = 0.96 + 1.04*(number of dry months [>100 mm rainfall]) + 0.21*(community size)

% Move time = 102.14 − 0.11*(annual rainfall, mm) + 0.00003* (annual rainfall, mm)2 + 2.597 * (foraging party size)

%Foraging party size = 21.49 + 0.07*(forest cover, %) − 0.33*(mean monthly rainfall, mm) + 0.001*(mean monthly rainfall, mm)^2

% Enforced rest time = − 29.47 + 1.28* (mean annual temperature, °C) + 0.34 * (% leaf in diet) + 5.95* $T_{mo}sd$

% Social time = 1.55 + 0.23* (community size)

Maximum tolerable community size can then be calculated, for any given location, as: (100 − (Feed + Move + Rest) − 1.55)/0.23.

Rest time and Social time are generic equations derived from data for all primates. Feed and Move time equations are specific to chimpanzees.

CHAPTER 4: THE FIRST TRANSITION: THE AUSTRALOPITHECINES

1. 'Taxon' (plural: 'taxa') is a general term used by biologists to refer to a class of animals. Depending on context, it might refer to a species (e.g. the common chimpanzee, *Pan troglodytes*), a genus (the chimpanzees as a group, genus *Pan*) or even a biological family (e.g. apes in general, including the chimpanzees, gorillas and orang utans).

2. Average temperatures decline by 1°C for every 100 m increase in altitude, and by about the same amount for every degree of latitude away from the tropics.

3. Succulents are a diverse family of plants that typically have fleshy leaves, stems and roots, which allow them to store water in drier habitats. Among members of this family are the irises, orchids, the asparagus family and cactuses.

4. Members of the horse family, including zebras and asses.

CHAPTER 5: THE SECOND TRANSITION: EARLY *HOMO*

1. In fact, *H. erectus* in Asia never produced handaxes at all. Handaxes are found in profusion to the west and south of what has become known as the Movius Line (after the archaeologist Hallam Movius, who first pointed this out), which runs roughly from Scandinavia to the Bay of Bengal. Only crude stone tools known as *choppers* are found to the east of this line, perhaps because of a lack of suitable stone. It is always possible that, in Asia, *erectus* made more sophisticated tools out of bamboo and other hard plant material, and that these have just not survived.

2. Species often become dwarfed on islands, especially when there are few natural predators.

3. The feeding time requirement is adjusted by the ratio of the metabolic body weights of *Homo* to that of the australopiths. Metabolic body weight (which is calculated as the 0.75 power of body mass, or $M^{0.75}$) indexes the relative energy consumption of body tissue. Metabolic body mass is also adjusted for the differential energetic costs of brain and somatic tissue. For the calculations, I have used the average body mass of the two sexes.

4. Two criticisms have been levelled against the expensive tissue hypothesis. One was that there is no trade-off between brain size and some measure of gut size in either South American monkeys (Allen and Kay 2012) or

mammals as a whole (Navarette et al. 2011), so the hypothesis must be wrong. This is rather an odd conclusion, because the expensive tissue hypothesis was an explanation that specifically addressed a problem in human evolution – how humans broke through a critical energy constraint. The whole point is that it *is* a ceiling, and hence the constraint only kicks in once brain size exceeds that seen in great apes. That it doesn't apply in birds or small-brained South American primates is neither interesting nor relevant. That said, an ingenious experiment on guppies (a small European fish that bears live young) has shown that, when artificially bred for increased brain size (and, as it happened, improved cognitive abilities, at least in females), the fish compensated by reducing gut size, as well as reducing litter size (the well-known fertility trade-off) (Kotrschal et al. 2013). The second criticism is potentially more serious. Isler and van Schaik (2012) argued that, because lifehistory processes have to be slowed down to accommodate larger brains, there is an inevitable point at which breeding rates become too slow for populations to maintain themselves, and that this creates a grey ceiling just above the brain size of great apes beyond which it is not possible for an ape to increase brain size without driving itself to extinction because it cannot breed fast enough to replace natural mortality. They argued that hominins broke through this grey ceiling by adopting communal breeding, which allowed a radical shortening of the interbirth interval (with help, mothers could afford to gestate a second baby while still lactating for the first and this helped raise fertility above the replacement rate). Unfortunately, the data cited to support this claim don't in fact show that birth rates would drop below the maintenance level: the relationship between brain size and fertility is not linear, as was claimed, but rather asymptotic (rather than falling below the replacement rate it bottoms out just above it). Even if we use the regression line to predict natural human interbirth intervals rather than use current human natural interbirth intervals, humans are still above the limit. The mistake arises from confusing a linear relationship on a double-log plot (i.e. an exponential [asymptotic] relationship) with a linear relationship in raw data.

5. In the past, palaeoanthropologists have frequently argued that the loss of large ape-like canines in the australopithecines and of large molars in early *Homo* was a consequence of the fact that they were no longer required for the kinds of diet these species had adopted or because they were able to use tools to fulfil the feeding functions for which these teeth had been necessary. However, in evolutionary terms, traits cannot be lost just

because they are not needed; they can be lost only because there is explicit selection against them or because they become incompatible with the mechanical requirements of another trait that is under positive selection. Contrary to what many have assumed, large canines are not used in feeding in primates (unlike carnivores), but are purely weapons used in male–male conflict. Loss of large canines certainly implies that direct competition between males was reduced, and this might well have been necessary to allow many males to associate together in large foraging parties; an alternative explanation might be that large canines get in the way of free jaw movement during chewing, which might have become disadvantageous under the more pressured foraging conditions of the australopiths. The reduction in the size of the molar tooth row in early *Homo* (such as it was) could have been to allow a reduction in jaw size for improved control over the articulatory space for vocal communication (e.g. laughter) as much
as a change in diet. If the diet was the driving force, then we need to show that there was a significant mechanical advantage or disadvantage in large cheek teeth rather than simply the loss of need. It is not obvious what that advantage or disadvantage might have been in respect of food processing.

6. Also known as the Turkana Boy, since the fossil was found on the edge of Lake Turkana in northern Kenya; he is also known by his Kenya National Museum accession number WT 15000.

7. Fifty-five per cent of the diet yields nutrients at the standard rate, and the rest (45 per cent) yields 50 per cent more nutrients, giving an overall improvement of 55% + (45% × 1.5) = 122.5 per cent.

8. It was claimed that there were burned sediments indicating a hearth at the famous Zhoukoudian caves in northern China, occupied by *H. erectus* between 0.75 and 0.2 million years ago; however, these have since turned out to be just discoloured soil.

9. The predators that co-occur with gelada are mainly leopard and hyenas (though lions were also present historically). Although leopards are now rare in the Ethiopian highlands, hyenas are extremely common (and the Ethiopian subspecies is unusually large and ferocious).

CHAPTER 6: THE THIRD TRANSITION: ARCHAIC HUMANS

1. Oxygen comes in two isotopes (^{16}O and ^{18}O), one slightly heavier than the other. Because the heavier isotope is more likely to be deposited as

snow than the lighter one when the climate is colder, the ratio of the two isotopes in an ice core (or in the skeletons of plankton) provides an index of temperature (the ^{18}O index).

2. The three Dryas events that occurred at this time involved short but dramatic periods of cooling, variously attributed to nuclear-winter effects from major volcanic eruptions (the Lacher See volcano in Germany was very active at this time), a large meteor impact in North America or shifts in the North Atlantic Conveyer (the current that brings warm water into the northern hemisphere from the tropics).

3. A clade is a group of species descended from a common ancestor species.

4. There were some precursor species associated with this breakaway group, notably *Homo rhodesiensis* in Africa and *Homo antecessor* in Spain, but we can ignore the taxonomic details and focus on *Homo heidelbergensis* as the archetypal archaic humans.

5. Late Neanderthals have been associated with a more sophisticated industry known as the Châtelperronian (after the site of Grotte des Fées at Châtelperron in southern France where it was first discovered) that used the new Levallois techniques of stone knapping (see note 8 below). However, not everyone is convinced that the makers of Châtelperronian tools included Neanderthals.

6. Calculated, as previously, with a maximum of 45 per cent of the diet devoted to meat and tubers (based on modern hunter-gatherers) at 50 per cent additional nutrient extraction, which would reduce this component from 45 per cent of feeding time costs to 45/1.5 = 30 per cent, for a saving of 44−30 = 14.0 percentage points.

7. The isotopes of nitrogen and carbon in collagen derive directly from the foods we eat. Because of the way chemical elements are incorporated into bone, an animal's nitrogen level is always a fixed amount higher than the amount in its food, and this allows us to tell what it had eaten over the last few years before it died. The carbon isotope levels allow us to distinguish their prey species according to whether they inhabited marine, terrestrial or freshwater environments.

8. The Levallois technique (named after the French site of Levallois-Perret, now a suburb of Paris) is a distinctive type of stone knapping developed during the later Palaeolithic. It involved a more sophisticated method of lithic reduction in which flakes were struck from a prepared core by creating a trimmed striking platform at one end. When the striking platform is finally hit, a flake separates from the core with a distinctive

convex profile whose edges had been sharpened during the initial trimming phase. This method provides much greater control over the size and shape of the final flake than previous knapping techniques. The flakes were sometimes adapted to produce projectile points known as Levallois points for use in spears and arrows.

9. This may be because at high latitudes goats, being essentially tropical animals, have to seek shelter in caves at night in order to cope with low night-time temperatures (Dunbar and Shi 2013).

10. Pearce et al. (2013) give slightly larger values because (at the insistence of the paper's referees) they recalculated the social brain relationship against cranial volumes rather than estimated neocortex size. While this still yields a significant relationship, it introduces more error variance by including irrelevant brain regions (brain stem, midbrain, cerebellum, etc.) and thus inevitably making the slope shallower, and this in turn reduces the difference between the species.

CHAPTER 7: THE FOURTH TRANSITION: MODERN HUMANS

1. For better or for worse, in the process they managed to wipe out the entire American mega-fauna, including giant species of mammoths, mastodons, sabre-tooth cats, horses and giant sloths among many others.

2. They speak languages that use a variety of clicks made with the tongue as consonants. These languages are thought to be very ancient. Clicks have been lost from all other descendent language families.

3. With a 50 per cent increase in digestibility from cooking, 12 percentage points of feeding would give the equivalent of $12 \times 1.5 = 18$ per cent of time spent feeding. Since brain size and body size predict that modern humans should spend 88.5 per cent of their day feeding, the 18 per cent represents $18/88.5 = 20$ per cent of the total feeding time (or, equivalently, of the diet – assuming, as we have done, that time spent feeding is roughly equivalent to proportion of the diet).

4. Although the limit on conversation group size is four, language can obviously be used for giving lectures to much larger numbers – but *only* providing the audience agrees to abide by cultural rules that allow the lecturer to hold the floor uninterrupted. Of course, we already have to have language in order to be able to agree to those cultural rules.

5. Modern humans are predominantly right handed (and, according to the archaeological record, always have been: the Nariokotome Boy, dated to 1.5 million years ago, seems to have been right handed). Handedness in fossil hominins can be determined from the angle at which the knapper strikes the stone when making tools (Cashmore et al. 2008).

6. Aiello (1996) pointed out that this includes a long litany of necessary changes: an ape-like flat chest, a bipedal gait to free the chest wall muscles off from being locked during locomotion, control over breathing and the articulatory space, and a lowered larynx, all of which are of ancient origin – but all of which had to be, successively, in place before language could evolve.

7. The hypoglossal canal claim has been challenged on the grounds that *some* australopithecine individuals have values that exceed those of *some* modern human individuals, and that this overlap therefore implies that these particular australopithecines had language (DaGusta et al. 1999). It should be obvious that this claim is specious, for several reasons: (1) it confuses variances with means, and it is means that determine species' competences; (2) it ignores the fact that a single individual with enough control to produce speech sounds does not make a language community, and language *is* after all a social phenomenon (there has to be more than one of you capable of speaking); and (3) it overlooks the fact that the thoracic nerve, hyoid bone and ear canal data all tell essentially the same story, thus providing some degree of corroboration.

8. At least in precocial mammals like anthropoid primates. Altricial mammals (such as rodents and many carnivores) give birth to premature babies that complete functional development in a nest.

9. Dunsworth et al. (2012) recently claimed that there is no obstetric constraint and that the problem really lies in maternal energy costs: birth occurs in humans when the mother reaches the point where she cannot sustain the energy cost of the growing baby any more. Unfortunately, this explanation completely ignores the fact that fuelling the baby's growth outside the womb through lactation is approximately 1.7 times more costly to the mother than doing so internally. This is because lactation, despite being an excellent solution to allowing an extended infancy, is a very inefficient use of energy: approximately 40 per cent of energy taken in for lactation is lost in the process of converting raw energy and protein into milk. If there really was an energetic constraint, it would have been much more efficient to complete the 21 months' gestation in the womb and find an alternative solution to the wide hips problem.

10. Perikymata represent growth lines in the enamel of teeth. Teeth grow by laying down successive layers of enamel, much as trees do in creating 'tree rings'. Each layer corresponds to about a week's growth. The layers appear as a series of lines (known as the striae of Retzius) on the surface of the tooth root that are visible under the microscope.

11. Symbionts are organisms that establish a mutually beneficial co-existence. They might be two species that live together in some way, or a micro-organism that takes up residence within a host where it provides a service for the host (as in the case of our many intestinal bacteria). A pathogen, in contrast, is a micro-organism that is detrimental, or even lethal, to its host (as in the bacteria and viruses responsible for many illnesses, including measles, typhoid and malaria).

12. There was a major series of eruptions in the Caucasus that did affect central Europe, though this actually happened around 40,000 years ago, so may even have predated the arrival of AMH in Europe, never mind the demise of the Neanderthals some 10,000 years later. Conceivably, given the timing, it might have been these eruptions that forced AMH to move westwards into Europe from the Russian steppes.

13. One of the problems, perhaps, has been a perceived need to explain what Neanderthals were doing with brains as big as those of modern humans, given the reasonable assumption that it was these big brains in AMH that created the cultural explosion of the Upper Palaeolithic Revolution. But, as I showed in Chapter 6, their *functional* brains weren't as big as those of modern humans because so much of it was given over to visual processing. Cognitively, they were still archaic humans, and there isn't anything to explain.

14. It has been argued that the human frontal lobe is no larger, relatively speaking, than those of other monkeys and apes: when the frontal lobe is regressed on total brain volume, humans lie on the general primate regression line (Semendeferi et al. 1997; Barton and Venditti 2013). However, the regression line is on a double-log plot with a slope of approximately 1.2, which means that the frontal lobe in fact gets relatively bigger as the brain gets bigger, and does so exponentially. In any case, since it is absolute brain volume, not relative brain volume, that determines neuron number, and hence cognitive competences, it wouldn't make any difference if the slope was unity since humans would still have absolutely bigger frontal lobes than monkeys and apes.

15. Many of the early immigrants to Iceland were men who had been banished from Scandinavia or England, and thus likely to have gone without

accompanying womenfolk. The genetic evidence suggests that a stop-off on the way to pick up some women was a common practice.

16. We know from the historical records exactly what happened: whenever the Mongols took a city that had resisted them, they slaughtered all the men while the women were, to put it politely, absorbed temporarily into the Mongol entourage.

CHAPTER 8: HOW KINSHIP, LANGUAGE AND CULTURE CAME TO BE

1. Spear-throwers are essentially sticks with a notched end into which the butt-end of a spear can be inserted, in effect extending the length of the arm when throwing. This allows greater angular momentum during the throw, resulting in the spear travelling at much greater speeds and for considerably longer distances than when using just the arm on its own.

2. The six are: English (as spoken in the UK and the USA), Laland (or Lowland) Scots (the traditional language of non-Gaelic southern Scotland), Caribbean English, Black Urban Vernacular (the language of urban African Americans), Sierra Leone creole and New Guinea pidgin. These differ in significant ways in both the words they use and their grammatical structures.

3. Parallel cousins are the offspring of a parent's same-sex siblings, cross cousins those of a parent's opposite-sex siblings.

4. Kinship is reckoned unilaterally when it is calculated only through one parent, and bilaterally when it is reckoned through both parents. English reckons kinship bilaterally (we use the terms *cousin*, *aunt* and *uncle* for both sides) but some languages, like Gaelic, use different terms for maternal and paternal relatives.

5. Matrilineal descent systems, in which relatedness is reckoned through the female line, seem to be mainly associated with cultures in which paternity certainty is low (i.e. males can't be sure that their wives' children were actually sired by them) (Hughes 1988).

6. The reason is that genetic relatedness declines very quickly the further back one goes in a pedigree: one doesn't have to go back much more than half a dozen generations for everyone to be more or less equally related to everyone else, especially in small communities. Hughes (1988) argued that

pedigrees are an ingenious solution to the fact that, while you and I should calculate our relatedness to the current pubertal generation (i.e. the cohort that is just about to reproduce), this cohort changes year by year, making it difficult to have a stable hook on which to hang our relatedness to each other. Hughes showed that referring back to a distant ancestor provides a fixed stable point that we can all relate to. More importantly, it turns out that it doesn't really matter whether this ancestor is real or mythical: you can include the man-in-the-moon in your pedigree and so long as he is far enough back it makes absolutely no difference at all to how well two people are related.

7. Hamilton (1964) identified kin selection as a major evolutionary force: if I help a brother or sister reproduce more successfully, then whatever genes I share in common with them that underpin this behaviour will be transmitted into the next generation just as if I had reproduced myself. Hamilton used this insight to explain the evolution of altruism towards kin, but the concept has since come to play a foundational role in modern evolutionary biology. In general, one should be willing to help relatives in proportion to their degree of relatedness to oneself, something that can be calculated rather precisely from knowing whose offspring they are.

8. Royalty and the nobility typically trace their ancestors back much further, but they are special cases: their rights to land or privilege usually depend on inheritance, and their long pedigrees usually exist to legitimate their control over these rights.

9. Some (e.g. Boyer 2001) have claimed that there are no health benefits to religion, but this seems to fly in the face of the evidence, which suggests that a broad range of benefits derive from being actively religious (Koenig and Cohen 2002). It may be that there is some confusion here over whether it is the religion itself (*qua* belief in God) or active involvement in religion and rituals that is the issue. It is almost certainly the latter that is important rather than the former. Merely claiming to be a member of a religion does not guarantee any effects and if this is what Boyer and other have focused on, then it is not surprising they found no effect. Cognitive science of religion seems to focus almost to exclusion on the more sophisticated concepts associated with doctrinal religions rather than on the highly emotional experiential aspects of religion that lie at the real heart of the religious experience.

CHAPTER 9: THE FIFTH TRANSITION: THE NEOLITHIC AND BEYOND

1. The Epipalaeolithic covers the first 1,000 years after the end of the last Ice Age.

2. Punishment in this sense is also known as second order altruism: the punishing of freeriders is done by altruistic members of society who are prepared to pay a cost for the greater benefit of society.

3. Naltrexone locks on to the μ-receptor sites for β-endorphins and effectively neutralizes them by preventing the endorphins accessing the sites. Since naltrexone is entirely neutral in its effect, it blocks the opiate high and mild analgesia that endorphins would normally produce, leaving only the negative side of alcohol.

4. One might point to religions like Buddhism that don't have formal gods as a counter-example. However, Buddhism is, in fact, not all that different: there is a set of standards of behaviour (usually at the recommendation of acknowledged holy men) and the individual is enjoined to monitor his or her adherence to these (subject, of course, to the impersonal accounting procedures of the Wheel of Karma, from which one's behaviour in life cannot be hidden).

5. This does not necessarily mean that one person can be simultaneously in love with several other people to the same degree. But it does mean that romantic relationships exist even within polygamous and polyandrous marriages. In these cases, they may well vary in intensity between the various male–female pairings and may be sequential rather than simultaneous for any one individual.

6. This is a consequence of sperm competition. When many males are likely to mate with a female, the male who leaves the largest volume of sperm in the female's reproductive tract is most likely to sire the offspring. Hence, in promiscuous mating systems, there is selection pressure for males to be able to produce increasingly large quantities of sperm, necessitating proportionately larger testes (Harcourt et al. 1981). Even though gorilla males may have to fight to acquire and maintain control over their harem, once ensconced they have exclusive mating access to all the females in their group, and so do not need to invest in producing large, relatively expensive testes.

7. The fact that not all individuals in a given culture experience the phenomenon we refer to as 'falling in love' or that the intensity of this effect varies between individuals as well as from culture to culture does not obviate the fact that it occurs in all cultures and throughout history (for

reviews, see Jankowiak and Fischer 1992, Fisher and Marcus 2006, Dunbar 2012) and is thus a human universal.

8. This has recently been suggested by Lukas and Clutton-Brock (2013), but the suggestion that hominin females ever foraged alone in large individual territories is clearly nonsensical; more generally, their rather simple analyses of the primate data fly in the face of the more sophisticated analyses of Shultz et al. (2012) and Opie et al. (2013).

9. Large prey may provide a lot of meat, but they are much harder to get than smaller prey, and produce poorer returns than gathering vegetable foods in terms of the time and effort invested. Their real merit is that they are dangerous, and so an uncheatable test of a male's quality and courage (and hence his genes).

10. The claim that this pattern is due to the fact that human females didn't normally live beyond the age of 35 years prior to the last century ignores the fact that it is the age at death of *reproductive* females (i.e. those that make it through puberty) that counts in evolutionary terms, not the age at death of everyone born (which undoubtedly was as low as 30 years in many Palaeolithic and historical populations). Women who survive to puberty commonly make it to 60 in both hunter-gatherer and historical populations, as a quick visit to old churchyards will confirm.

11. Or if their day journeys were one-tenth the length of the typical chimpanzee day journey (5 km). This seems very unlikely: it would imply that australopithecine day journeys were less than 0.5 km, and that seems implausibly short (it isn't even true for baboons living in riverine habitats, whose day journeys are typically 1–2 km).

12. In other words, one in which the females are individually bonded to the harem male, but the male is not especially bonded to the females.

13. If males can keep rivals completely away from their females by defending an exclusive territory, then they can continue to monopolize a number of females. But there is a limit to the extent to which they can do this: once the number of females exceeds about a dozen, the number of excluded males becomes so large that it is difficult for the harem-holder to prevent other males invading his group (Andelman 1986, Dunbar 1988).

14. Unfortunately, Pérusse (1993) defined monogamous males as, in effect, those in a marriage. This will inevitably inflate the proportion of strictly monogamous males because, at any given time, the number of married males will, in Western cultural environments, inevitably include some promiscuous males in a temporary marital state.

15. The Sexual Orientation Index, or SOI (Penke and Asendorpf 2008).

Index

References to tables and figures are indicated by italic page numbers only when these appear outside the pages or page ranges where the topic is discussed in the text.

Topics from the endnotes after page 383 are distinguished by 'n' and the note number as in 384n2